DIESELS and ELECTRICS ON SHED

(Volume Four)

SCOTTISH REGION

DIESELS and ELECTRICS ON SHED

(Volume Four)

SCOTTISH REGION

by Rex Kennedy

Oxford Publishing Co. Oxford

ISBN 0 86093 043 2

Negatives & Plates by:
Novascan Ltd., High Wycombe

ACKNOWLEDGEMENTS

My thanks go to all who have contributed photographs for inclusion in this album, and especial thanks go to my friend, Tom Noble, who lives near Glasgow, for his help with information on workings within the Scottish Region of British Rail. Sincere thanks also go to my wife, Pauline, who has now spent two holidays with me in Scotland, where the beautiful mountain scenery and picturesque lochs were often replaced by locomotive depots and stabling points.

Class 55 'Deltic' locomotives were not introduced onto British Railways until 1961, and at first carried no nameplates. Standing nameless, outside the old steam depot at Haymarket, in November 1964, is No. D9010. This locomotive was later named *The King's Own Scottish Borderer.*

Barry Wynne

Half title page
A Metro Cammell d.m.u. sits at the entrance to Dundee depot on 11th August 1979. The picture clearly shows the type of brickwork used at this particular Scottish location, which differs greatly from other depots on this region.

David Maxey

Frontispiece
Scotland has more than its fair share of snow, which, for the railway photographer makes an ideal setting. On New Year's Eve 1978, Class 20, No. 20 080 stands proudly, in the snow, at Eastfield depot, before commencing yet another year of hard work in 1979.

Tom Noble

Title page
Grangemouth depot was originally established by the Caledonian Railway. On 14th October 1978, a solitary Class 37 locomotive, No. 37 155, protrudes from the depot entrance. This class of locomotive is one of the more common forms of motive power used at this depot for freight haulage.

Tom Noble

Printed by:
B. H. Blackwell Ltd. in the City of Oxford

Published by:
Oxford Publishing Co. Ltd.,
8 The Roundway,
Headington, Oxford

INTRODUCTION

The Scottish Region of British Rail has seen, since the introduction of diesel traction north of the border, a variety of classes from the Class 17 'Claytons' and North British Class 21 and 29 locomotives, to the Class 47/7 Brush locomotives, fitted for push-pull working, and the new Class 140 two-car diesel multiple unit. Since electrification to Glasgow and Coatbridge, depots, such as Polmadie, now stable Class 81–87 locomotives, where ex-LMS 'Duchesses' and 'Black 5s' once stood in preparation to work trains from Glasgow (Central), southbound, via the West Coast Main Line.

Many of the depots, such as Parkhead, Bathgate, Perth (South) and Mallaig have now sadly disappeared from the Scottish scene, but in this volume, the fourth in the 'on shed' series, I have tried to include as many as possible of these locations. Locomotives and units now seem to be stabled wherever they complete their last turn of duty.

Clydeside is now a mass of electrified lines carrying Class 303, 311 and 314 electric multiple units and offers an excellent service to commuters in the Glasgow area. Class 26 and 27 locomotives are still very prolific on the Scottish Region and more and more d.m.u. routes, such as Inverness to Aberdeen, Edinburgh to Dundee and Glasgow (Queen Street) to Oban have now become locomotive hauled services. Class 47/7 locomotives now operate the Edinburgh (Waverley) to Glasgow (Queen Street) service, once under the control of the converted Class 27s operating in pairs, one at each end of the train.

The Scottish Region was the last stronghold of the Class 06 shunter but since their withdrawal from operating stock, only Class 08 shunters remain.

Visits by enthusiasts to Scotland, where they can now only see five or six major depots, are a far off cry from years ago, when the region was scattered with interesting engine sheds.

New depots have sprung up in Scotland over the years, such as Hyndland and Glasgow (Shields Road), the latter housing and maintaining the new overhead electric Advanced Passenger Train which entered regular service between Glasgow and London (Euston) in December 1981.

We shall soon note the disappearance of the Class 55 'Deltic' from the Edinburgh area in 1982, and as the Class 25 and Class 40 locomotives reduce in numbers every year, the variety of traction in Scotland will soon get less and less.

In this volume I have tried to include as many locations and classes of locomotive as possible, and in an effort to make this book interesting, have coupled the pictures with captions containing detailed information and a little history.

I hope that this album, portraying Diesels and Electrics on Shed in the Scottish Region, will perhaps bring back a few memories for those who remember the days when the variety of classes was greater, and also interest those enthusiasts not having had the opportunity of visiting the region.

Rex Kennedy
Oxford 1982

There could be no more appropriate locomotive to accompany the introduction to a book covering the Scottish Region of British Rail, than the one named *Saint Andrew*, after the patron saint of Scotland. Class 47/7, No. 47 701, bearing this name, stands in Millerhill yard on 27th April 1980.

Rex Kennedy

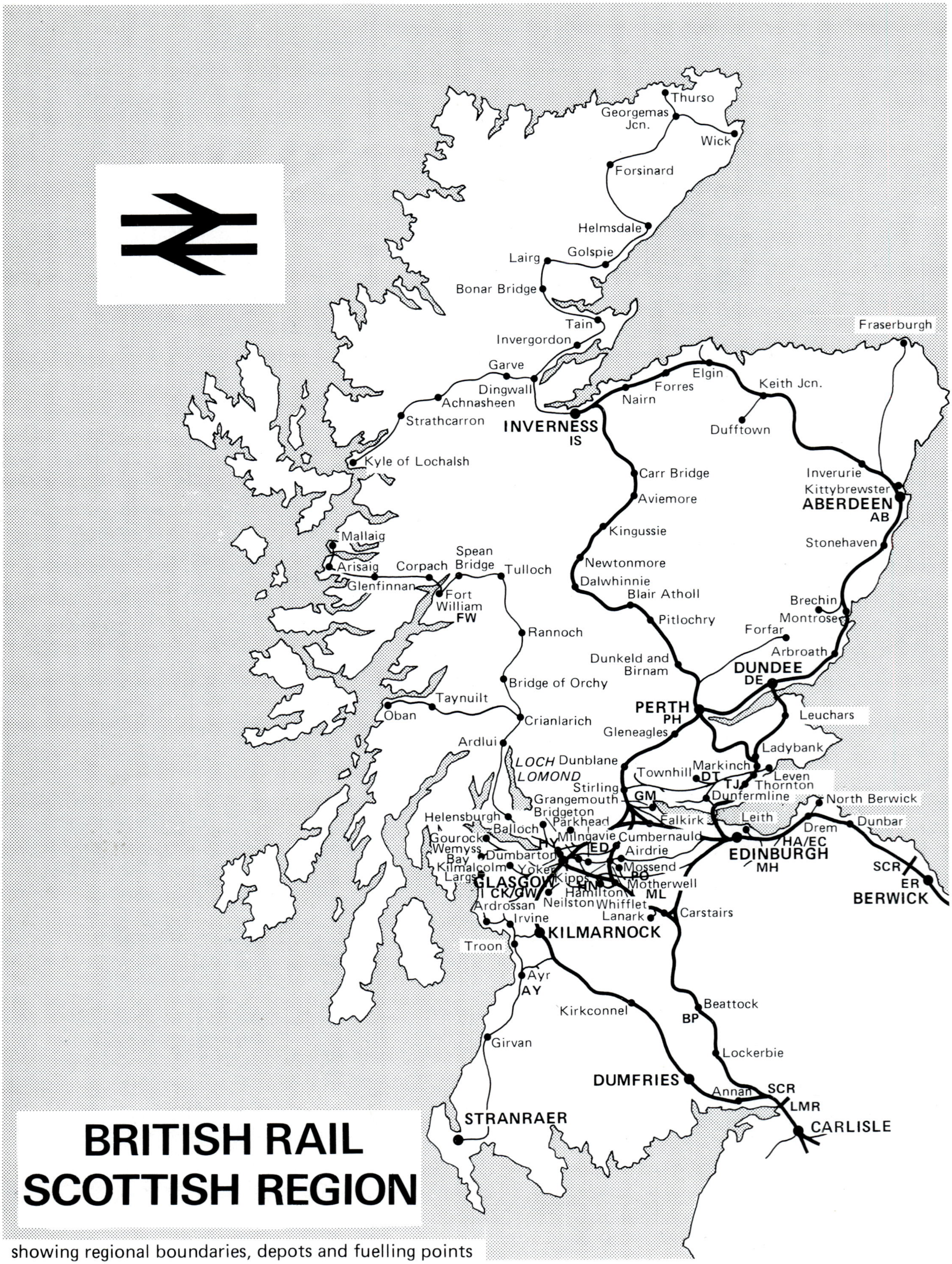

BRITISH RAIL SCOTTISH REGION

showing regional boundaries, depots and fuelling points

BRITISH RAILWAYS MOTIVE POWER DEPOTS, FUELLING POINTS and STABLING POINTS of the SCOTTISH REGION portrayed in this volume

Location	*Code*	*Old Codes*	*Plates*
Inverness Maintenance Depot	IS	32A/60A	1–13
Kyle of Lochalsh Stabling Point	–	32A (Sub)/60A (Sub)	14
Wick Stabling Point	–	32A (Sub)/60D	15
Aviemore Depot	–	32B/60B	16–18
Kittybrewster Depot	–	61A	19–22
Aberdeen (Ferryhill) Servicing Depot	AB	29B/61B	23–35
Aberdeen (Waterloo Goods) Stabling Point	–	–	36
Keith Junction Depot & Stabling Point	–	61C	37–38
Elgin Depot & Stabling Point	–	61C (Sub)	39–40
Fraserburgh Stabling Point	–	61A (Sub)	41
Thornton Junction Depot & Fuelling Point	TJ	62A	42–46
Dundee D.M.U. Depot	DE	29C/62B	47–54
Montrose Stabling Point	–	62B (Sub)	55
Townhill Servicing Depot	DT	–	56–61
Dunfermline Depot	–	62C	62
Markinch Stabling Point	–	–	63
Kirkcaldy Stabling Point	–	62A (Sub)	64
Perth (South) Depot	–	29A/63A	65–67
Perth Station Fuelling Point	PH	–	68–73
Stirling Stabling Point	–	31B/63B	74
Fort William Depot & Fuelling Point	FW	63B/63D/65H/65J	75–81
Oban Stabling Point	–	31C/63C/63D/63E	82
Edinburgh (St. Margaret's) Depot	–	64A	83–85
Millerhill Fuelling Point	MH	–	86–94
Haymarket Maintenance Depot	HA	64B	95–116
Edinburgh (Craigentinny) HST Depot	EC	–	117–118
Dalry Road Depot	–	28B/64C	119–120
Leith Central Depot	–	64H	121
Eastfield (Glasgow) Maintenance Depot	ED	65A	122–142
St. Rollox Depot	–	31A/65B	143
Whifflet Junction Stabling Point	–	–	144
Parkhead Depot	–	65C	145–146
Yoker Stabling Point	–	31E (Sub)/65G	147
Gourock Stabling Point	–	–	148
Grangemouth Depot	GM	31D/65F	149–154
Polmadie Servicing Depot	PO	27A/66A	155–174
Glasgow (Shields Road) Maintenance Depot	GW	–	175–183
Hyndland E.M.U. Depot	HY	–	184–185
Bridgeton Central E.M.U. Stabling Point	–	–	186
Balloch Central E.M.U. Stabling Point	–	65I	187
Airdrie E.M.U. Stabling Point	–	–	188
Helensburgh Central E.M.U. Stabling Point	–	65H	189
Milngavie E.M.U. Stabling Point	–	–	190
Motherwell Fuelling Point	ML	28A/66B	191–201
Hamilton D.M.U. Depot	HN	66C	202–209
Mossend Yard Stabling Point	–	–	210–211
Carstairs Depot	–	228C/64D/66E	212–215
Beattock Stabling Point	BP	66F/68D	216
Corkerhill Servicing Depot	CK	30A/67A	217–224
Ayr Maintenance Depot	AY	30D/67C	225–235
Irvine Stabling Point	–	–	236
Stranraer Stabling Point	–	67F/68C	237
Ardrossan Depot	–	30C/67D	238–239
Ardrossan Harbour Stabling Point	–	–	240
Dumfries Depot & Stabling Point	–	67E/68D	241–243

INVERNESS

The old depot

2 The depot at Inverness has sadly changed since the days of steam and transition, losing the well known landmark, the arched water tower, pictured here. The old steam depot also pictured, was actually situated on the opposite side of the main line to the present diesel depot. In this view, Class 26, No. D5330, sets the scene of transition with Class 5 ex-LMS 4-6-0 No. 45117, complete with snowploughs, and the turntable pit in the foreground. *D. Porter*

3 Another view of the old Inverness depot, this time showing the impressive coaling tower in the background. Varying styles of early green liveries, carried by Class 26 diesel locomotives, are seen in this picture, taken on 17th August 1960. A BR 2-6-4 tank locomotive is also in attendance. *John Hooper*

4 The turntable overlooked by the roundhouse at Inverness is another sight from the old steam depot built in 1862. On 13th August 1960, Class 26, No. D5345, was seen being rotated, with the single line token catcher clearly on view at the far end of the locomotive. The roundhouse had twenty-one bays, three of which led to the repair shop, and two being through roads to the perimeter of the shed building. The turntable was built at a cost of £320 in 1862. *Peter Groom*

◀ **1** Inverness, now the most northerly depot on the Scottish Region, has, in recent years, provided us with a predominance of Class 26 locomotives. This view, as seen on 16th August 1981, shows two Class 26 diesels, Nos. 26 021 and 26 033. The front locomotive, No. 26 021 is fitted with snowploughs, a familiar sight in Scotland, but carries only one centrally situated headlight, whereas these locomotives are normally fitted with either two, or none at all. *Tom Noble*

Lochgorm

5 The interior of Inverness (Lochgorm) Works, pictured here, is now part of the present day diesel depot, and has changed very little since the early days of diesel traction in Scotland. This photograph shows Class 26 locomotives, with the hinged headcode discs in varying positions, lined up over the inspection pits.

D. Porter

6 A rare sight at Inverness is that of a Metro-Cammell unit, as seen in this picture, undergoing maintenance in the depot on 12th June 1978. This type of multiple unit, built mainly for short distance services, would not have been used on the only d.m.u. service from Inverness to Aberdeen. One can only assume that it had arrived from Perth for major repair. Class 25, No. 25 087 and a Class 47 diesel stand alongside on the next road.

Tom Noble

7 This old stone building, originally part of the old Inverness (Lochgorm) Works which stood on this site, regularly has a Class 26 locomotive poised for a picture at its entrance. With the reflection of the warning panel of the shed door on the side of the locomotive, No. 26 015 starts her engine in preparation to move off on 20th April 1980, whilst No. 26 042 awaits a 'wash and brush-up' in the background.

Rex Kennedy

8 A general view of the depot not only portrays the variety of main line diesels to be found here, but also the contrast of stone and brick buildings now to be seen at Inverness, the stone building again being part of the old Lochgorm Works built in 1861. Locomotives pictured on 12th June 1978 are Class 47, No. 47 517 (HA), Class 25, No. 25 044, and moving off the fuelling point, Class 26, No. 26 034.

Tom Noble

Refuelling

9 The depot fuelling point at Inverness, stands close to the line which bypasses the station for trains from the far north of Scotland to the south. Class 26, No. 26 040 takes on fuel on 3rd June 1978. This locomotive is fitted with multiple working jumper leads to facilitate double heading, a feature of Class 26 working in Scotland.

John Augustson

11 A fine example of an ex-Highland Railway signal box stands between the roads that lead into the depot at Inverness, and the main line into the station from the south. On 2nd August 1978, Class 37, No. 37 219 stands beside this box. ▶

Graham Scott-Lowe

10 Whilst Class 20, No. 20 102 is being refuelled, the cab windows receive a polish before she returns to duty. This locomotive, together with No. 20 080, seen behind, has just arrived at Inverness with a freight from Elgin, on 5th September 1979.

Michael Rhodes

12 The depot area at Inverness is clearly visible from the station, and sits in a triangle bound by the line to the far north of Scotland, the main line to the south and the station avoiding line. With a view of the offices in the background, two Class 20 locomotives, both newly outshopped from St. Rollox Works, Nos. 20 204 and 20 125, stand off duty, whilst in the station, Class 26, No. 26 032 prepares to leave with the 18.05 train to the Kyle of Lochalsh on 29th May 1981.

Les Nixon

13 A three car Class 120 Swindon cross country d.m.u. headed by power car No. SC51793, refuels at Inverness depot prior to working another service to Aberdeen on 23rd January 1980. Class 26 locomotive No. 26 043 stands alongside. The Class 120 multiple units were used, until 14th April 1980, on trains to Aberdeen, a service since taken over by ex-Eastfield Class 27 locomotives. The majority of these units were sent to Wales after the changeover.

Brian Morrison

KYLE OF LOCHALSH

14 All that remained of the depot building at the Kyle of Lochalsh in 1967 was a small retaining wall. On the 21st August of that year, Class 24, No. D5115 stands at this location, before returning with a train to Inverness.

Roger Lamb

WICK S.P.

15 Trains from Inverness to Wick and Thurso, in the far north of Scotland, split at Georgemas Junction where a spare locomotive waits, and the two smaller trains continue to their separate destinations. Although a depot building once stood at Wick in the days of steam, locomotives are now stabled in the station. On 13th February 1979, Class 26/1 No. 26 040 waits in the station, being held overnight, in preparation to leave for Inverness with the 05.30 train on the following day, a journey of 161¼ miles.

Brian Morrison

AVIEMORE

16 The depot building at Aviemore is now used by the Strathspey Railway, a preservation group, running steam hauled trains from their own platform at Aviemore to Boat of Garten. In 1960, steam and diesel could be seen together here, and this view shows a Class 26 diesel locomotive outside the depot on 13th August of that year, with an ex-LMS Class 4, Fowler 2-6-4 tank engine, simmering away. A Class 20 diesel can also be seen inside the depot building.

Peter Groom

17 The type of stone often used when building the Highland Railway depots, is shown in the view taken at Aviemore in the 1960s. Class 26, No. D5340, fitted with snowploughs and single line token catcher, basks in the sun outside the depot, whilst ex-LMS Fowler 2-6-4T No. 42168 waits behind the office building.

D. Porter

18 Four-wheeled Park Royal Railbus No. M79971, fitted with a B.U.T. (A.E.C.) 6 cylinder horizontal type 150 bhp engine, stands at Aviemore depot on 11th November 1961. This vehicle was used on the Speyside line, (ex-Great North of Scotland Railway), from Aviemore to Boat of Garten, Grantown on Spey, Craigellachie and Elgin. The picture indicates that at this time the railbus still carried the 'M' prefix to its number, a reminder of the days when it was used on the Bedford to Hitchin branch, a link route between London Midland and Eastern Regions.

Graham Wise

KITTYBREWSTER

19 Kittybrewster depot, now closed, along with the station of the same name, was situated just north of Aberdeen, and in 1947, a tram service operated between Aberdeen and Kittybrewster stations. This view of the unusual half round-house at Kittybrewster, taken on 9th August 1966, shows North British Class 21 locomotives together with diesel shunters inside the depot.

David Wood

20 In the dark depths of Kittybrewster roundhouse lies Class 06 Barclay 0-4-0 shunter No. D2421. This particular class of locomotive was introduced onto British Railways in 1958, specifically for use on the Scottish Region.

J. Coltas

21 North British Class 21 diesel electric locomotive, No. D6150, stands in the yard at Kittybrewster on 19th July 1961. The 'North British' works plates on the cab side are clearly visible.

John Oatway

22 Preparing to carry out shunting duties in the yard at Kittybrewster, Class 06 shunter, No. D2422, waits with staff 'at the ready', to move off, on 19th July 1961.

John Oatway

23 A variety of diesel traction is lit by the bright sunlight at Aberdeen (Ferryhill) in September 1977. Motive power includes Classes 47, 08, 26, 25 and 40. The buildings have changed very little over the years, apart from alterations to the roof of the main depot. *Les Nixon*

24 Standing at the fuelling point at Ferryhill on 16th September 1978, are two Class 26 diesel locomotives, Nos. 26 010 and 26 012. With them stands a Class 40. Amongst other locomotives on view are Class 47, No. 47 270 and three Ferryhill allocated Class 08 shunters, Nos. 08 828, 08 817 and 08 710. Class 40, No. 40 142 is seen in the distance moving on to the depot from the direction of Aberdeen station.

Tom Noble

25 The first of three views portraying the changes at Ferryhill, over the years, shows the water columns still standing in the yard outside the depot entrance, in April 1965. Class 26 diesel locomotives still mingled with the odd steam locomotive around this time.

David Thornley

26 By September 1967, the water columns had disappeared at Ferryhill, but the steam vents still remained on the roof of the shed building, although many of the roof windows had been covered in. On this occasion, a solitary Class 47, in the two tone green livery much loved by enthusiasts, stands in the shed yard.

Roy Jones

The changing face of Ferryhill

27 Eventually, all roof windows were disposed of at Ferryhill depot and the roof vents, from the days of steam, had also disappeared. On 16th September 1978, the depot yard housed Class 40, No. 40 029 *Saxonia* without any form of head-code panel, No. 40 148 with single headcode panel and Class 25, No. 25 246. Also in attendance was Class 122 Gloucester R. C. & W. Co. single car unit No. SC55002, allocated to Hamilton depot, and bearing the Greater Glasgow Passenger Transport Executive symbol.

Tom Noble

Refuelling

28 By 1981, Class 47 locomotives were more prolific at Ferryhill depot than the Class 40 diesels. On 22nd July 1981, No. 47 120 refuels with Nos. 47 520, 47 424 and 47 528 in close proximity. Class 26, No. 26 026, complete with snowploughs, stands at the entrance to the maintenance shed.

Bert Wynn

29 The mighty power of two Class 40 locomotives at the fuelling bay at Ferryhill, is portrayed in this view of Nos. 40 163 and 40 165 on 8th December 1974. *Brian Morrison*

30 The interior of the maintenance depot at Ferryhill, seen on 10th August 1979, shows Class 25, No. 25 239 standing over the inspection pits. This particular locomotive is one of those fitted with roof headcode box and horns each side of this box.

David Maxey

The maintenance shop

33 Class 55 'Deltic' locomotives were quite a common ▶ sight at Aberdeen before the introduction of the High Speed Train service from Kings Cross to Edinburgh and Aberdeen. Since then, visits by 'Deltics' have been less frequent. On 21st February 1979, No. 55 005 *The Prince of Wales's Own Regiment of Yorkshire*, bearing one of the longer names in the class, stood in Ferryhill yard.

John Chalcraft

31 The instructions on the 'spangle' fitted to Class 47, No. 47 207 standing inside the maintenance depot at Ferryhill, speak for themselves, as the locomotive undergoes major repair on 16th September 1978.

Tom Noble

▲
32 A pair of Class 40 locomotives, Nos. 40 162 and 40 166 stand at the buffers in Ferryhill yard on 10th August 1974. Again the differing headcode panel designs are distinctive, with No. 40 166 only carrying the single panel.

Rex Kennedy

34 Ferryhill's solitary Class 06 shunter, No. 2423, later renumbered 06 006, stood in the depot yard, amidst wagon wheels, on 10th August 1974. She is pictured here in her usual immaculate condition, and could regularly be seen operating with the steam crane, in the shed area, prior to her withdrawal.

Rex Kennedy

ABERDEEN (Waterloo Goods)

KEITH JUNCTION

◀ **35** A view looking out from the maintenance depot at Ferryhill, on 8th December 1974, shows two Class 40 locomotives, Nos. 40 152 and 40 142 with differing headcode boxes. Caught by the sunlight, Class 47 No. 47 470 moves off the depot in preparation to take the 12.20 (Sunday service) Aberdeen to Kings Cross express.

Brian Morrison

◀ **36** Waterloo Goods Yard is situated in the dock area of Aberdeen, and a shunter diagram is allocated to this location from Ferryhill depot. The Class 08 shunter, on this occasion No. 08 828, stood here on 28th August 1978 during an off duty period.

R. E. Ruffell

37 Keith Junction is situated on the Inverness to Aberdeen line at the point where the, now freight only, line runs south to Dufftown. In steam days, the line to Dufftown continued on to Craigellachie, the junction for Elgin and Aviemore, but now these lines no longer exist. This view of the old steam depot shows a variety of steam traction, together with a North British Class 21 diesel electric locomotive. Both steam and diesel locomotives appear well equipped for severe winter weather conditions.

D. Porter

38 Only a small part of the old steam depot now remains at Keith, and a Class 25 locomotive can often be found stabled here, after running grain from Aberdeen to the distillery at Dufftown, which is along the line seen disappearing into the distance in this picture. No. 25 233 is viewed here, stabled, on 18th April 1980.

Rex Kennedy

ELGIN

39 The end of another era. Stored North British Class 21 locomotives, including Nos. D6144, D6143 and D6134, lie dormant inside the old steam depot at Elgin, awaiting their fate, in September 1967.

Roy Jones

40 With the steam depot at Elgin now gone, and a new station built on the site of the old Highland Railway station, locomotives can often be found stabled near to the old Great North of Scotland Railway station, in the yard. This old station is still a fine building, and it would be a shame to see it disappear. Before departing with a freight, Class 25 locomotives, Nos. 25 109 and 25 229 wait in the yard on 10th August 1979.

David Maxey

FRASERBURGH

41 This view of Fraserburgh, portraying the grandeur of the buildings overlooking the freight yard and station, shows Class 26, No. 26 029, as it lies stabled before commencing shunting duties, after a typical Buchan cloudburst, on 29th January 1979. At this time, the only outward traffic on the branch was parcels and seed potatoes. After much local opposition, the branch closed on 6th October 1979.

George O'Hara

THORNTON JUNCTION

42 The 'Kingdom of Fife' was once solely North British Railway territory, and Thornton Junction once sported a fine steam depot. This night view shows steam and diesel together during the last months of its existence, with the sinister silhouette of the coaling tower looming high above the depot. A Class 40 diesel and an 0-6-0 J38 ex-LNER locomotive share the generous accommodation at Easter 1966.

Les Nixon

The shed

43 Hunslet 0-6-0 Class 06 diesel shunter No. D2583 stands in the yard outside Thornton Junction steam shed on 22nd May 1959, together with ex-LNER Class J37 0-6-0. No. 64565 and J38 No. 65909 in the background.

Norman Preedy

44 Final days at Thornton Junction steam depot with Class 17 'Clayton' locomotive No. D8581 and another member of the class, and Class 40 No. D358 standing outside the depot building. The unusual roof design of this shed is an interesting feature. This depot was situated east of the present diesel fuelling point near the A92, south of the line, and opposite Rothes pit, now closed after flooding.

John Chalcraft

Fuelling point

45 This general view of Thornton Junction, as it stands today, shows the open site where locomotives and multiple units are stabled and refuelled, with the hills of Fife beyond. A Class 101 Metro-Cammell unit, headed by power car No. SC51245, lies stabled in the yard, whilst a Class 27 locomotive stands near the fuelling point, on 14th October 1978. Class 08 shunter No. 08 341 waits to continue its yard duties beyond the array of point levers in this busy yard.

Tom Noble

46 Class 20, No. 20 110 stands off duty near the fuelling point at Thornton Junction on 17th June 1981. Thornton Yard is used mainly for the marshalling of coal trains from the Fife coalfields. The line, which goes away to the right of the picture under the road bridge, is to Westfield pit, near Ballingry.

Rex Kennedy

DUNDEE

47 This view of Dundee depot shows everything familiar to this Scottish location. The depot structure, which has changed very little since this picture was taken on 24th July 1961, the River Tay, the Tay Railway Bridge, with the sight of a steam-hauled freight train approaching Dundee, and the presence of a d.m.u. and small diesel shunter in the yard. The shunter, North British 0-4-0 No. 11710 awaits refuelling, whilst a Metro-Cammell unit completes the scene.

Courtesy British Rail

48 A more recent view of Dundee depot, taken on 18th October 1978, depicting a Class 27 locomotive, No. 27 016, with Metro-Cammell power car No. SC50158 from a Class 101 two-car multiple unit set, a class introduced on to British Railways in 1956.

Tom Noble

49 An interior view of the depot at Dundee, with main line traction, commuter transport, and a locally allocated shunter standing over the inspection pits on 11th August 1979. Metro-Cammell units are a familiar sight at Dundee, and this particular two-car set stands alongside Class 08 shunter, No. 08 712 and Class 27, No. 27 021.

David Maxey

50 Overlooked by the spire of the church, which stands beside the depot yard at Dundee, Class 26, No. 26 019 rests alongside the refuelling bay on 16th November 1979.

Alan Clarke

Shunters at Dundee

51 North British Co. 0-4-0 shunter No. 11710 sporting the 'lion on wheel' British Railways crest and makers plate, shunts at Dundee depot yard on 19th July 1961. The cab design on these locomotives ensures good visibility for the crew.

John Oatway

52 Class 06 shunters stand back to back over the pits inside Dundee depot on 10th January 1981. No. 06 002, with all yellow front panel, was used for duties over the tight curves on Dundee Harbour branch, a line since closed.

John Augustson

53 Two withdrawn Class 06 shunters, Nos. 06 006 and 06 005 await their fate in Dundee depot yard on 30th March 1981. Alongside, is the sole survivor of the class, No. 06 002, which has now, sadly, been withdrawn. These three particular members of the class all had three windows on the rear of the cab, where some only carried two.

Tom Noble

54 The turntable, a relic from the days of steam at Dundee, creates an interesting foreground for a Class 25 and Class 27 diesel, No. 27 016 on 18th October 1978. The common sight of a Metro-Cammell d.m.u. completes the scene. The wooden slats, forming the footholds from the past, for manual rotation of the turntable, add further interest. Dundee depot had the threat of closure hanging over it in December 1981, following the change in services from Dundee to Edinburgh in October 1981 from d.m.u. to locomotive hauled. Since this action, most of the multiple units have been transferred to Ayr depot.

Tom Noble

55 Locomotives are occasionally stabled in the bay at Montrose station. On 21st April 1980, Class 40 locomotive, No. 40 066 lay here jacked up due to an axle problem, and later in the week, was to be found in Dunfermline (Townhill) depot, undergoing repair.

Rex Kennedy

56 Dunfermline (Townhill) depot and wagon works is a ▶ centre for freight locomotives, mainly Class 20s. Wheel and brake repairs are carried out in this depot, and often even a d.m.u. is to be found here under repair. On 27th April 1980, No. 20 225 undergoes maintenance alongside wagons.

Rex Kennedy

MONTROSE S.P.

DUNFERMLINE
(Townhill)

57 On 22nd March 1980, a row of five Class 20 locomotives stand inside Townhill depot. No. 20 217 heads Nos. 20 191, 20 222, 20 203 and 20 216. These locomotives work m.g.r. and coal trains in the area.
Graham Scott-Lowe

58 A class of locomotive, now sadly missed by diesel enthusiasts is the '24'. This class was introduced in 1958, and on 14th June 1975, No. 24 069 was seen standing in the yard at Townhill.
Norman Preedy

59 Class 20 locomotive, No. 20 191 waits beside the fuelling point at Dunfermline (Townhill) depot, on 27th April 1980. On the Midland and Eastern regions, the Class 20 locomotives regularly work in pairs, but in Scotland they are often seen alone, hauling short freight trains. The Class 25 alongside is No. 25 010.

Rex Kennedy

60 This view of Townhill, showing the depot structure and the segregation between the wagon section of this location and the locomotive area, on 27th April 1980, portrays views of both end features of Class 20 diesels. Nos. 20 219 and 20 221 stand side by side, whilst a Class 25 locomotive is stabled in the distance.

Rex Kennedy

61 Two locally allocated Class 08 shunters, Nos. 08 726 and 08 175, stand with the remains of Class 06 shunter, No. 06 008, in Townhill yard on 27th April 1980. The 06 locomotive had been cannibalised in an effort to keep the remaining members of this small class in running order. A Class 08 shunter from this depot is often loaned out to the N.C.B. at Comrie Colliery nearby.
Rex Kennedy

DUNFERMLINE

62 Steam and diesel were seen together in 1960 at the old steam depot at Dunfermline. On 12th June of that year, two North British Co. 0-4-0 diesel shunters stood near the coaling tower and, uniquely, one carried the original number 11717, and the other, the revised number, D2707. Another interesting feature is that the yellow and black markings on D2707 differ from those on an engine of the same class, in **Plate 51**, at Dundee. Dunfermline depot was completely re-roofed with glass in 1955/56, which was rather unusual for a steam depot, with the dimensions of the building measuring 270 ft x 68 ft, and housing four tracks.
F. Hornby

64 Until recent years, a Class 06 shunter could also be found stabled at Kirkcaldy station at the end of the platform beneath the signal box. The signal box has now been taken away and the entire class of 06 shunters withdrawn. On 14th June 1981, Class 08 shunter, No. 08 441 stood in Kirkcaldy station yard stabled with a brake van.
Rex Kennedy

MARKINCH S.P.

63 A Townhill allocated Class 06 shunter, until withdrawal of this class from operating stock, could always be found at Markinch. This locomotive was required for work on the sidings to two paper mills, now all that remains of the Leslie branch, and it could usually be found, at weekends, stabled beside the shed near the station, which is seen in the background of this picture. On 28th June 1980, No. 06 003 stands down the yard, with Class 20, No. 20 204 passing light engine. Two weeks after this picture was taken, the panel box at Edinburgh took over signalling operation of this area. This particular Class 06 shunter can now be seen at Reading, on the Western Region, bearing the number 97 804, as a departmental locomotive, having replaced the well-known Ruston & Hornsby 0-4-0, No. 20 at Reading Signal Works.

Michael Rhodes

KIRKCALDY S.P.

PERTH (South)

65 A 'Tesco' supermarket now stands on the site of the, now demolished, steam depot at Perth. As late as 1969, diesels could be found at this location, and this general view, taken in July 1966, shows a Class 26 diesel locomotive moving off the depot with little else around. Ex-LNER A4 No. 60024 *Kingfisher* can be seen disappearing into the distance with an Aberdeen to Glasgow express.

John Hooper

67 A Class 20 locomotive basks in the sunshine at the old Perth steam depot location, and makes a fine locomotive study. It is noticeable that this particular member of the class carries no headcode box. Discs, as seen here, were fitted instead to Nos. 20 001–20 128.

David Thornley

66 The transition period at the old Perth steam depot at Easter 1966. Ex-LMS Black 5s, Nos. 44722 and 44797 guard the depot with ex-LNER A4, No. 60009 *Union of South Africa*. BR Pacific locomotive No. 72006 *Clan Mackenzie,* simmers away on the right. These locomotives, together with a Class 27 diesel, create the scene. A fine selection of British Rail motive power.

Les Nixon

68 Refuelling at Perth station, a location now used as a major stabling point in Scotland. The local hillside overlooks this fine station, and the high wall hides other locomotives stabled beyond. In this view, on 27th May 1981, two Class 40 locomotives, Nos. 40 158 and 40 173 prepare for their next duty.

Les Nixon

PERTH STATION S.P.

69 An ex-works Class 25 locomotive, No. 25 086, stands in the bay, used for the stabling of diesels, at the north end of Perth station, on 24th April 1980. This bay is now fenced off from the rest of the station and the fuelling point is also located in this vicinity.

Rex Kennedy

70 This view, taken from the north end of Perth station, showing the Highland line to Inverness in the foreground, portrays a Class 27 locomotive, No. 27 020, and a Class 26, No. 26 019 standing near the fuelling point, on 26th June 1981. The fine structure of Perth station can be seen in the background. It was once called Perth General and, until 1923, was owned jointly by the Caledonian, Highland and North British Railway Companies.

Rex Kennedy

71 The high wall at Perth station which often separates stabled locomotives from the main line through the station, has this archway, which forms an ideal frame for the composition of an unusual picture. On 14th November 1973, after diesel locomotives had lost the 'D' prefix, a Class 24/2, No. 5129 lay stabled with a Class 50 locomotive, No. 429, prior to the 50s being transferred, in their entirety, to the Western Region. A two-tone green Class 47, No. 1579, from Stratford depot on the Eastern Region, completes the scene.

Brian Morrison

72 A view from the other side of the wall at Perth stabling point shows a Dundee allocated Class 08 shunter, No. 08 725, off duty, on 3rd April 1976, with a glimpse of the station through the archway. The Class 08 shunter, used as Perth station pilot, now actually bears the Perth town crest on the back of the cab, and its duties include the shunting of wagons at the motorail terminal beside the station.

Tom Noble

73 A final check before moving off for Class 47 locomotive, No. 47 546, as it stands beside the fuelling bay at the north end of Perth station, on 24th April 1980. This class of locomotive is now a common sight on the Highland line to Inverness.

Rex Kennedy

74 A Class 47 diesel can often be found lying stabled at Stirling station, which is overlooked by the fine turreted buildings of the town. This beautifully kept station is an impressive example of Caledonian Railway architecture, although it was completely rebuilt in 1914, in an updated version of the traditional stepped gable style. On 20th June 1981, No. 47 008 stands in the Stirling sunshine.

STIRLING S.P.

Rex Kennedy

75 The new diesel depot at Fort William lies beneath the shadow of Ben Nevis, and locomotives for use on the West Highland line to Glasgow, and also to Mallaig, are stabled there. Locomotives off timber trains from Crianlarich to the, now closed, Pulp Mill at Corpach nearby, no longer appear here since the withdrawal of this service. On 25th July 1981, the scene was predominantly Class 37s, with No. 37 051 in the foreground, and a sight of a Class 37 bearing the new front end livery, now a feature of the area since the introduction, in June 1981, of Class 37 locomotives on passenger trains on the West Highland line. Class 20 locomotive, No. 20 002 and an 08 shunter, No. 08 443 complete the picture. At Glasgow (Queen Street), on 6th October 1981, four Class 37 locomotives were named. These were No. 37 043 *Loch Lomond,* No. 37 081 *Loch Long*, No. 37 026 *Loch Awe* and No. 37 027 *Loch Eil*. These are four of twelve of the class to be named for use on passenger trains on the West Highland line, and all four named carry a West Highland White terrier dog painted beneath the nameplate.

Graham Wise

37051

76 One has to go a long way back to remember the old North British Railway steam shed at Fort William, complete with its roof. This depot was not situated on the same site as the new diesel depot, but diesels were stabled near to the deteriorating shed building as late as the early 1970s. The shed bordered Loch Linnhe, and lay between the old Fort William station, situated in the town beside the loch, and Mallaig Junction. This view, on 21st August 1967, shows two Class 29 locomotives, one inside the depot building, and the other in the yard, alongside. Behind the shed building lay the turntable, the pit for this having been cut out of solid rock. The new station at Fort William is, in fact, built near the site of the old steam shed.

Roger Lamb

77 This overall view of the area around the diesel depot at Fort William, shows how it nestles in the mountains, situated just off the line to Mallaig. On this summer day, 20th June 1977, Class 25, No. 25 108 lies amidst the tranquillity of this highland scene.

Tom Noble

78 As black clouds hover over Fort William depot, Class 20 locomotive, No. 20 089 awaits refuelling on 1st September 1978, before taking on its next task, which will be freight back to Cadder Yard, Glasgow, now only a permanent way yard since its closure as a marshalling yard in December 1980.

John Dunford

79 Class 25/1 locomotive, No. 25 079, pulls up to the doors at Fort William diesel depot after coming off duty on a sunny 15th February 1979. Sitting beside the depot in the shadows, is Class 27/0 locomotive, No. 27 023, complete with snowploughs.

Brian Morrison

80 Two Class 27 locomotives, a class regularly used on passenger trains from Glasgow (Queen Street) to Fort William before the introduction of Class 37s, stand outside the only building with roller doors at Fort William depot. Wagons and permanent way equipment are usually stored in the open shed seen to the left. Nos. 27 202 and 27 042 are pictured here on 22nd April 1980.

Rex Kennedy

81 The open-sided building at Fort William depot housed, on Sunday, 9th August 1981, a Class 20 locomotive No. 20 002, and Class 08 shunter No. 08 443. Both these locomotives were, at that time, operating shunting duties at Fort William, the 08 at the Corpach Paper Mill and the Class 20 working as station pilot. A Class 20 locomotive is used at Fort William for this purpose, instead of an 08 shunter, as it is powerful enough to haul a passenger train, should there be a failure.

Terry Harper

82 Locomotives are only stabled at Oban, the end of the Caledonian route to the west coast of Scotland, pending return to Glasgow on passenger or freight services. This view, on 16th May 1977, shows this fine Caledonian station, built in 1880, with many of the platforms no longer being used since the closure of the line to Ballachulish, on 28th March 1966, and the reduction of services to and from Glasgow. Class 25, No. 25 235, between freight duties, lies stabled, as Nos. 25 087 and 27 003 prepare to leave with a train for Glasgow (Queen Street) later in the day. Owing to the sparse timetable at Oban, locomotives often wait on their trains for some considerable time, before moving off.

Tom Noble

OBAN S.P.

EDINBURGH (St. Margaret's)

83 The ex-North British Railway depot at Edinburgh St. Margaret's has sadly gone, missed by all who still like to see diesel locomotives in the environment of the old steam shed. It was situated about 1½ miles east of Waverley station on both sides of the main line. Its roundhouse was on the north side of the line, but its main shed, on the south side. This view, taken in November 1965, shows North British Co. 0-4-0 diesel shunter, No. D2715 being turned manually on the turntable at the depot.

Les Elsey

84 A different type of North British 0-4-0 diesel shunter, No. D2706, moves wagons on the depot yard at Edinburgh St. Margaret's on 2nd August 1962, amongst the still apparent 'steam' atmosphere. This locomotive carried the Davey Paxman 200hp engine, differing from the one in the previous picture, which carried the North British 225hp engine.

Peter Groom

85 A feature of St. Margaret's depot were the North British shunters, and this scene, on 20th July 1961, shows yet another member of the same class as that pictured in **Plate 83**, as it stands over the pits, on one of the lines which form the turntable network. It is particularly noticeable on No. D2745, that the yellow and black markings, on the front of the locomotive, differ from others in this class, like the shunter behind, and the one seen in **Plate 83**. The British Rail crest on the locomotives also differs in design.

John Oatway

86 Millerhill depot is situated in the vicinity of the East Coast Main Line, but cannot be seen by passengers travelling over that route, as it lies in the network of suburban freight lines, south of Edinburgh. The landmark, which overlooks the yard at Millerhill, is that of Monktonhall Colliery, which can be seen from some distance away. A large variety of motive power can be found here, and this view, on 17th April 1978, shows Classes 26, 40, 47 and an 08 shunter on shed.

Tom Noble

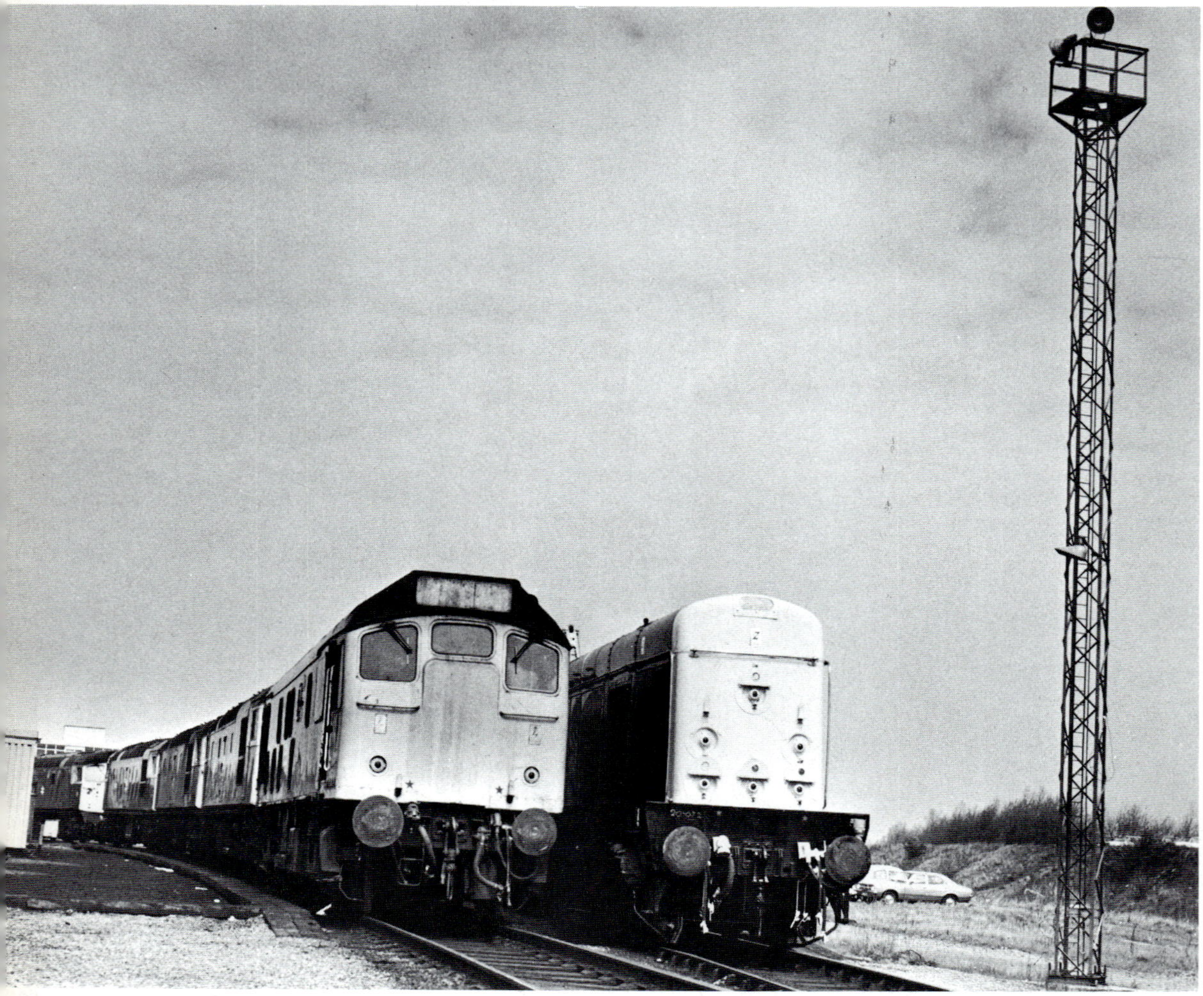

87 At weekends, two lines of stabled locomotives usually lie beside the only depot building at Millerhill. These vary in power from Class 20 upwards. On 27th April 1980, one complete line of Class 20s was headed by No. 20 072, whilst the other line included Class 25 and 26 locomotives, Nos. 25 068, 26 021, 26 002, 26 003 and 26 001. Class 20s are used, double headed, from the nearby Monktonhall Colliery, and amongst their other duties, they take m.g.r. trains to Longannet Power Station.

Rex Kennedy

88 With the large British Rail office block overlooking the depot and yard at Millerhill, a solitary Class 26 locomotive, No. 26 007, stands beside the depot on 17th April 1978. A Class 08 shunter, allocated to Haymarket, like all the shunters that operate the yard, lies stabled in the distance.

Tom Noble

89 Class 40 locomotive No. 40 033 *Empress of England* pulls out from the depot building at Millerhill on 18th June 1981. Fuelling facilities can be seen beside the depot entrance. This particular locomotive is one of the many dual braked Class 40s, and is still fitted with centre communication door, but no headcode panels. Headcode discs have been removed, and the six marker lights seen, include the ones that once lay beneath those discs.

Rex Kennedy

90 In the mid and late 1960s, Class 17 'Clayton' locomotives were a common sight at Millerhill. These 900bhp locomotives, powered by two Davey Paxman six cylinder horizontal engines, were introduced on to the Scottish Region in 1962, mainly for use on short distance freight, or local passenger trains. This view, on 7th June 1965, shows five members of this class, with No. D8557, fitted with snowploughs, in the foreground.

Martin Welch

91 A later view of a Class 17 'Clayton' locomotive at Millerhill, differing very little in paint design from the one in the previous picture, apart from around the cab windows. No. D8587 is seen here, on 8th September 1971, and together with No. D8586, was fitted with Rolls-Royce engines.

Norman Preedy

92 Class 40 locomotive, No. 368, bearing its original number before the new T.O.P.S. numbering system was introduced in 1973, stands in Millerhill yard on 8th September 1971. This was one of the later built locomotives of the class, fitted with central headcode box and having no centre communication door. It is noticeable that the previous picture, taken on the same day, portraying the 'Clayton' locomotive, showed the 'D' prefix on the number and the 'lion' crest, whereas, on this locomotive, the 'D' has been removed, and it also bears the new 'arrow' emblem.

Norman Preedy

93 Over the years, 'Peak' Class 46 locomotives have been a fairly common sight at Millerhill, but as this class seems to be slowly diminishing, this may not be for much longer. On 1st October 1979, No. 46 046 stood beside the depot with Class 26 locomotive, No. 26 008 inside.

Tom Noble

94 Five class 08 shunters stand at Millerhill on a wintry 22nd March 1980, headed by Nos. 08 561 and 08 571. This busy yard is a hive of activity on weekdays, and shunting operations, with freights constantly arriving and departing, are in force most of the time.

Graham Scott-Lowe

HAYMARKET

95 Haymarket depot is situated west of Edinburgh (Waverley) station, beyond the tunnels, Waverley Gardens and Haymarket station, and can be seen from the main line, from trains to and from the Forth Bridge and Glasgow. This very busy depot always provides a good variety of traction, and this general view of the depot, on 27th April 1980, showing the varying building designs, depicts two Class 47/7 locomotives, Nos. 47 706 *Strathclyde* and 47 704 *Dunedin* together with No. 47 219 and two Class 40 locomotives, with differing front-end design, Nos. 40 158 and 40 080.

Rex Kennedy

97 One of the maintenance workshops at Haymarket is seen in this picture, showing, on 20th August 1979, Class 25 locomotive No. 25 086, having been lifted from its bogies, whilst Class 26, No. 26 001 stands over the inspection pit, alongside. Class 26 locomotives, Nos. 26 001–26 007 are used on m.g.r. trains from Bilston Glen Colliery, Monktonhall Colliery and Blindwells Opencast Site, to Cockenzie Power Station. The very first m.g.r. trains, operated by British Rail, in fact, worked to Cockenzie Power Station, on its completion early in 1966. This power station was scheduled to use 3 million tons of coal per year, initially from Monktonhall Colliery, six miles away.

Tom Noble

98 Inside the same maintenance building, on 27th April 1980, stood two Class 40 diesels, Nos. 40 157 and 40 060. This class of locomotive is used in the area on Edinburgh to Aberdeen trains, Glasgow to Aberdeen trains, Edinburgh to Carstairs (WCML portion) trains, occasionally, and internal Scottish and Highland line freights.

Rex Kennedy

◀ **96** An earlier view of the Haymarket depot sees windscreen cleaning in progress, prior to the departure of Class 47 locomotive, No. 47 458 on its next turn of duty. Two Class 46 'Peaks', Nos. 46 046 and 46 043 complete the picture. Both 'Peaks' bear the headcode, 1E07, which indicates the 11.10hrs. service from Edinburgh (Waverley) to Kings Cross, and are pictured on 11th August 1974. It is most unusual to see a situation such as this, with two locomotives on shed, each bearing identical headcode, apart from the fact that it is also an unusual 'Peak' working.

Rex Kennedy

Deltics at Haymarket

99 A classic photograph from early diesel days. Resplendent in green livery, stand a 'Deltic', Class 40 and Class 26 locomotive, at Haymarket, when the old steam shed was still standing. Although each locomotive is painted green, the variety in paintwork design is very noticeable. The trio is made up of Nos. D363, D5310 and D9004, later to be named *Queen's Own Highlander.*

Eric Treacy

100 Two more early liveried 'Deltics' stand nosing out from the old steam depot at Haymarket. Resting over the pits is No. D9006, later named *The Fife & Forfar Yeomanry*, whilst its shed mate is No. D9009, later named *Alycidon*. It was four years before naming of the 'Deltic' class was complete. No. D9009 left the depot later to work the 'up' 'Flying Scotsman', and D9006, later that day, worked the 'up' 'Elizabethan'. A disused water column stands between the two locomotives.

Eric Treacy

Preparing for duty

101 Preparing for the off outside the old steam depot at Haymarket, and almost determined not to be outdone by the steam locomotives in residence, Class 55 'Deltic', No. D9006, looking fine in green livery, starts her powerful engines on 20th July 1961, not long after the introduction of this popular class of locomotive. She was later named *The Fife & Forfar Yeomanry,* but when naming of the class commenced in 1961, preference seemed, at first, to be given to the names of 'Derby' winners and not to regiments, the Eastern Region allocated 'Deltics' being the first to be named.

John Oatway

102 Class 55 'Deltic' No. 55 009 *Alycidon*, one of the earlier named locomotives in the class, roars away from Haymarket depot, to take out a Sunday working from Edinburgh (Waverley), on 11th August 1974. It is noticeable that other locomotives in the yard bear headcodes, whereas the 'Deltic' carries the two white marker lights in the route indicator panel. In fact, *Alycidon* was the very first locomotive on British Railways to have its headcode panel converted to marker lights.

Rex Kennedy

103 Another 'Deltic' scene at Haymarket, as No. 55 011 *The Royal Northumberland Fusiliers,* sends out a plume of fumes as it prepares to leave the depot on 3rd June 1978. The two large fuel tanks are situated between the bogies and can clearly be seen. This locomotive also sports the all yellow indicator panel with marker lights, differing from the 'Deltic' in **Plate 102**. Other locomotives on view include Class 47, Nos. 47 407, 47 268 and 47 408.

John Chalcraft

105 Diesel multiple units can often be found at the west end of Haymarket depot, outside the running shed. On 28th February 1978, a Metro-Cammell unit, still sporting the old route indicator panel, sits outside the depot. Multiple units are used extensively in the Edinburgh area, for journeys to Dunblane, North Berwick, Arbroath and into Fife. *Tom Noble*

106 A Class 105/2 Cravens motor brake second class power car, protrudes from the modern diesel depot at Haymarket on a spring Sunday morning, 27th April 1980. Edinburgh was the first city to have Inter-City units operating. These were Swindon Works Inter-City BR sets, and they started operating from Edinburgh to Glasgow in 1956.
Rex Kennedy

107 Multiple units are also stabled at the eastern end of the depot, together with diesel locomotives and various vans. On 27th April 1980, a Class 101/1 Metro-Cammell motor composite unit, No. SC50264, stands in this area with Class 25 and 26 locomotives, Nos. 25 033, 25 084 and 26 012. No. 25 064 is seen stabled in the distance.
Rex Kennedy

◀ **104** Portrayed in the foreground is the unavoidable greasy pit, which is familiar to any well-used maintenance depot. Standing outside, on 27th April 1980, is Class 20, No. 20 102. This class of locomotive is used for running double-headed m.g.r. trains from the Ayrshire coalfields to Longannet Power Station, and is also used on Fife coal trains and local freight and ballast trains.
Rex Kennedy

Class 47 – Liveries

108 Class 47 diesel No. D1510 stands in the forefront at the west end of Haymarket depot on 17th March 1963, whilst ex-LNER Class A1 4-6-2 Pacific, No. 60152 *Holyrood* sits behind, beside the water column. The diesel sports the fine two-tone green original livery of early BR days. The locomotive pictured here was one of the first twenty built at Brush Electrical, Loughborough, and destined for service on the Great Northern Line of the Eastern Region in 1962. The beginning of an era for the Class 47 diesel. *Dennis Anderson*

109 With a view of the fuelling depot in the background, Class 47 diesels, bearing the livery generally seen, are pictured at Haymarket depot on 15th August 1978. The locomotives on view are Nos. 47 417, 47 422 and 47 541. *Bert Wynn*

110 Nos. 47 711 *Greyfriars Bobby*, (ex 47 498), and 47 712 *Lady Diana Spencer*, (ex 47 505), are the two latest named additions to the Class 47/7 fleet, and they are seen here, together, on 28th June 1981 at Haymarket. This class of locomotive operates the push-pull service between Edinburgh and Glasgow (Queen Street), a service once under the control of Class 27/1 and 27/2 locomotives, operating in pairs with an engine at each end of the train. These two particular 47s, converted and renumbered in Crewe Works, ran un-named for about twelve months, and were originally intended to be named *William Wallace* and *Prince Charles Edward*. However, a competition to choose names, was launched by British Rail, in conjunction with the Glasgow and Edinburgh local radio stations, and the names eventually chosen were those now carried by these locomotives. The naming ceremony took place on 30th April 1981, No. 47 712, at Glasgow (Queen Street), followed later in the day by the naming of 47 711 at Edinburgh (Waverley). *John Augustson*

111 Class 47/7 locomotive, No. 47 703 *Saint Mungo*, emerges, sparkling, from the washing plant at Edinburgh (Haymarket) depot on 16th August 1980. The front end modifications, including centre headlight, are clearly visible. This particular locomotive was converted from Class 47/4, No. 47 514, for push-pull duties on the Glasgow to Edinburgh line, and reclassified 47/7.

M. J. Howarth

112 A trio of Class 27/2 locomotives stand outside the east end of the maintenance shop at Haymarket on an inclement day. This view portrays Nos. 27 206, 27 207 and 27 204 on 23rd November 1979. These locomotives were Class 27/1s converted from steam heating to electric train heating and reclassified 27/2.

Graham Scott-Lowe

114 Class 20, No. 20 175, stands beside the washing plant at Haymarket depot on 16th August 1980. Some of this class are fitted with both vacuum and air brakes, and many on the Scottish Region fall into this category. All the Class 20s used on m.g.r. trains are dual-braked and fitted with slow speed control.

M. J. Howarth

115 A very unusual visitor to the Scottish Region is the Class 31 locomotive. However, on 16th August 1980, No. 31 113 was to be found on Haymarket depot having worked up from the Eastern Region, possibly with a freight.

M. J. Howarth

116 A lone Class 26 locomotive, No. 26 012, sits in the sunshine outside Haymarket depot on 11th June 1975. This class of locomotive can regularly be seen working in pairs, and, invariably, the same two engines remain coupled together for a series of duties.

Norman Preedy

◀ **113** With a background showing Edinburgh Castle, a Class 40 locomotive, with disc head-codes and front communicating door, No. 40 069 stands at the east end of Haymarket depot on 13th August 1978. Since October 1981, no locomotives of this class were allocated to the Scottish Region.

John Hillmer

EDINBURGH (Craigentinny)

117 A carriage shed has existed for some years at Edinburgh (Craigentinny), where locomotives were often to be found. However, since the introduction of the High Speed Train to Scotland, this is now an HST depot. The depot is situated beside the East Coast Main Line, east of Edinburgh (Waverley) station. This interior view, on 1st September 1980, shows the duty Class 08 shunter, No. 08 881, over the pits, together with Class 47/7 No. 47 711, before the Class 47 was named, standing with Edinburgh to Glasgow push-pull coaching stock.

Tom Noble

118 Front '254' numbers now seem to have disappeared from Eastern and Scottish Region HST sets, but in this view, on 1st September 1980, No. 254 023 proceeds through the washing plant at Craigentinny, the second stage of the washing operation being the rinsing off of chemicals applied in the actual wash.

Tom Noble

DALRY ROAD

119 Dalry Road depot was an old Caledonian Railway depot, situated in the fork west of Edinburgh (Waverley) station, where the Glasgow line joined the line to Carstairs. A station of the same name also once existed, but now both depot and station are gone. In April 1965, Class 17 'Clayton' locomotives, as in this view, were a fairly regular sight at this location.

David Thornley

120 This side view of Class 17 'Clayton' locomotive, No. D8587, at Dalry Road, clearly shows the huge length of these 68 ton centrally cabbed locomotives. Unfortunately, their life span was comparatively short.

John Hooper

LEITH CENTRAL

121 It seems strange to see a depot, so large and full of motive power, as Leith Central, seen here on 6th January 1968, no longer operative. On this occasion, at least nine Class 08 shunters and two Class 05 'Hunslet' shunters stood inside the depot, together with multiple units under maintenance. The shunters were used in Leith Docks, and the only Class 05 shunter still working on British Rail is to be found on the Isle of Wight, now in departmental stock.

David Wood

EASTFIELD

122 This clear view of the structural design of the diesel maintenance depot at Eastfield, as seen from the west, shows the depot's all steel and glass building, with its familiar roof design, in its situation within the depths of urban Glasgow. We see a selection of Type 2 Bo-Bo locomotives on shed on 11th April 1980, including Nos. 27 042, 25 075 and 27 033.

Les Nixon

123 The east end of the depot at Eastfield, pictured here on 4th September 1977, shows the lay-out of this modern depot and its yard, together with the large variety of motive power to be found here. Classes seen in this view include 06, 08, 20, 25, 27, 37 and 47; a usual weekend scene.

Gavin Morrison

124 A general view of Eastfield, looking east, on 28th June 1981. Metro-Cammell units tick over, and Class 37s stand off duty. Type 2 locomotives stand on the far side of the depot complex. This view shows the entire width of the depot at Eastfield, clearly indicating the area used by the multiple units, and, to the left of the units, the old refuelling shed, no longer used for this purpose since the construction of a new fuelling point at the rear.

John Augustson

125 A typical modern traction scene at a busy motive power depot. As we, again, look at Eastfield from the east, Metro-Cammell and Gloucester R. C. & W. Co. units stand separated from the main line motive power in the form of Class 25, 27 and 47 locomotives, on 4th September 1977. Services operated by multiple units from Glasgow include those to Cumbernauld, Dunblane, and to Edinburgh via Falkirk Grahamston.

Gavin Morrison

126 A shaft of sunlight brightens the gloom inside the lifting shop at Eastfield on 26th December 1977. Class 27, No. 27 112, rests over the inspection pit during the festive season. This locomotive was one of the Class 27s fitted for push-pull working on the Glasgow to Edinburgh route.

Tom Noble

127 Class 25, No. 25 007, hangs in the air, suspended by the lifting equipment at Eastfield, on 15th October 1978. Class 20, No. 20 020 stands, with other locomotives, over the inspection pits alongside. A variety of parts and equipment completes the scene.

John Chalcraft

128 The refuelling building at Eastfield lies adjacent to the main line to Edinburgh, and on 15th October 1978, Class 25 locomotive, No. 25 026 stands awaiting fuelling before returning to duty. The later designed Class 25 locomotives carried the larger centre window at the end of the cab.

John Chalcraft

129 The familiar sight of the 'Trans-Clyde' symbol, carried on multiple units in the Glasgow area, is clearly visible in this view of a Metro-Cammell three-car set at Eastfield on 27th April 1980. It is noticeable that the d.m.u. maintenance building at this depot, is just long enough to accommodate a three-car set for underframe inspection and repair.

Rex Kennedy

Shunters at Eastfield

131 After the storm, Eastfield allocated Class 08 shunter, No. 08 738 stands on its home depot, reflected in the aftermath of the storm, as the weather favourably changes, on 19th August 1979. The Eastfield shunters' duties include Glasgow Works, Cowlairs carriage sidings nearby, and Stirling station, where it is station pilot. Until the closure of Sighthill yard, on 5th October 1981, a shunter also worked at this location.

Tom Noble

◄ **130** Dual braked Class 08 shunter, No. 08 938 propels four other members of its class at Eastfield depot on 27th March 1977, beside the Glasgow to Edinburgh main line. Other 08s in view, are Nos. 08 443, 08 721, 08 227 and 08 754. The only other dual braked shunter in this rake is No. 08 754.

Tom Noble

133 In July 1969, the present Eastfield depot was under construction, and as seen in this picture, the area in the background has already been cleared. This view, from the east end of the depot, portrays the old steam shed still in use, now taken over by diesel traction. Class 20s, a Class 40 and a Class 17 'Clayton' take pride of place from the ex-LNER B1s, N15s and K2s, which once stood here. The old steam shed was built in 1904, with space for 100 locomotives on its 14 roads.

Tom Noble

134 Standing outside the old brick built steam depot building at Eastfield, in ex-works condition, and bearing the 'lion' crest, is North British 0-4-0 shunter, No. D2755. This locomotive, pictured on 22nd May 1959, carries the modified radiator grille fitted to certain members of the class, a class of shunter very prolific on the Scottish Region during this period.

Norman Preedy

135 The Class 06 shunter is now sadly missed by the diesel enthusiast visiting Scotland. No. 06 004 could regularly be found in the early 1970s, at weekends, on Eastfield depot, in addition to the Class 24s which have also disappeared. On 11th August 1974, she was seen, with side panels removed, awaiting attention in the depot yard.

Rex Kennedy

◄ **132** Class 06 shunter, No. 06 003, photographed at Eastfield depot, on 29th June 1978, carries 206003, a very strange number. Unfortunately no explanation can be given for the figure '2' which precedes the T.O.P.S. number.

Graham Wise

136 Apart from the Class 24 locomotives now converted to heating units, like Nos. 24 054, 24 142 and 24 061, this last dying tribute to the class, No. 24 006, lay for some considerable time at Eastfield depot, unfortunately the victim of vandalisation and cannibalism. When this photograph was taken, on 19th August 1979, she had been resting there for over three years, but she has now gone forever.

Tom Noble

137 Class 25, No. 25 086, in ex-works condition, stands at Eastfield depot on 27th April 1980. This particular member of the class is of the later design with the larger centre window at the end of the cab, and incorporating the continuous front handrail. The complex underframe details stand out clearly.

Rex Kennedy

138 After the change of motive power from Class 27, on the Glasgow to Edinburgh push-pull service, in the late 1970s, twelve Class 47 locomotives were converted and reclassified 47/7 to operate this service. To enable only one locomotive per train to be used, as against two, when the Class 27s operated the route, Mk. III trailer driving cars were constructed to operate at the remote end of the train. The train's lighting cables were used to provide control to these cars. On 27th April 1980, No. 47 702 *Saint Cuthbert* stands on Eastfield depot along with Class 27, No. 27 034.

Rex Kennedy

139 Class 37 locomotive, No. 37 112, sporting the wrap-around yellow ends, is pictured at Eastfield depot on 31st August 1981. Together with No. 37 111, this locomotive is used on passenger services over the West Highland Line. The use of Class 37s over this route, which officially commenced in June 1981, reduced the journey time from Glasgow (Queen Street) to Fort William by some 20 minutes. A gradual infiltration of 37s had appeared on this service over a period of months prior to June 1981. Serious attempts to introduce this class started in 1978, when No. 37 108 regularly appeared over this picturesque route. No. 37 111 differs slightly from No. 37 112 in the yellow paintwork around the cab windows.

John Hillmer

140 Eastfield on 31st December 1978, as main line motive power in the form of Class 47, No. 47 541, stands at a snow-bound depot in the winter sunshine beside a Class 102 Metro-Cammell unit, headed by car No. SC51455. The crew of the d.m.u. prepare to start the engines of this frozen-up set in this picturesque view.

Tom Noble

142 Carrying the 65A shedplate beneath the number, Class 29 North British locomotive, No. D6103 lies at Eastfield depot in January 1967, at a time when, in addition to the sight of some of the new steel diesel depot buildings, signs of the steam days, like the coaling tower in the background, were still apparent. When these locomotives were built, all were Class 21, and carried 1,000 hp engines. However, from 1963–1967, twenty of the class were fitted with 1,350 hp engines and became Class 29. The majority of the Class 21s remaining, eventually received 1,100 hp engines. The only distinguishing feature externally between the Class 21 and 29 locomotive was the fitting of headcode windows at each end of the Class 29 on conversion.

Roy Jones

141 The tranquillity of a cold winter evening at Eastfield is portrayed here on 3rd January 1979. Overlooked by the multi-storey blocks of flats, pictured in this view, the depot is situated on the south side of the line from Edinburgh, before the line enters the tunnels that lead to Glasgow (Queen Street) station. The locomotives on this occasion, lying dormant, having completed the day's duties, are Nos. 47 269, 47 053, 25 082, 37 144 and 27 017 with snowploughs. An Eastfield allocated Class 08 shunter is also in attendance.

Tom Noble

ST. ROLLOX

143 The old steam depot at St. Rollox, Glasgow, open until the early 1960s, was situated in the same area as the famous ex-Caledonian Railway locomotive works. The works are, of course, still used today for diesel maintenance and overhaul purposes and can be found some distance away from the site of the old steam shed. During the transition period, steam and diesel could often be found, together, at the old steam depot, and on 12th August 1960, Class 26 diesel, No. D5344, immaculate in green livery, stands with an ex-LMS 'Black 5' in the background. The involved underframe details of the Class 26 locomotive are very clear.

Gavin Morrison

WHIFFLET JUNCTION S.P.

144 In the early 1970s, Class 06 shunters were a regular sight stabled near the bridge at Whifflet Junction, in the suburbs of Glasgow, and by 1974, when new T.O.P.S. numbers were being displayed, members of this class could be found at this location, withdrawn and still bearing their old numbers. In May 1972, No. D2413 lay stabled here beside the signal box.

Barry Nicolle

PARKHEAD

145 In steam days, Parkhead depot was used for the Glasgow suburban locomotives, and in 1965, a mixture of steam and diesel traction could be found at this location. This view, taken on 9th August 1965, looking through the shed, sees North British Class 21 locomotive, No. D6130, alongside ex-Highland Railway 4-4-0 'Small Ben' Class No. 54398 *Ben Alder*, which had been taken out of service in February 1953. *Ben Alder,* being the sole survivor, was stored for thirteen years, in the hope that she could be economically restored to full working order again, but this was not to be. She would have been a great attraction on the Strathspey Railway at Aviemore. No. D6130 was eventually uprated, with a 1,350hp engine, to Class 29.

Roger Lamb

146 North British Class 21 locomotive, No. D6125, stands outside Parkhead depot in April 1965. The North British diesel-electric locomotives, Class 21 and 29, spent many years, up to withdrawal, on the Scottish Region, and although ran trials on this region in 1959, were in fact, built for the Eastern Region, where the earlier built locomotives of the class spent their first days. In fact, Nos. D6130 and D6135–6157, ran on the Scottish Region from being built in 1960, at the time when earlier members of the class were being transferred to the region from the Eastern Region.

David Thornley

YOKER S.P.

147 Yoker, situated north-west of the city of Glasgow, was once the site of a sub-shed to Dawsholm (31E), in the ex-LMS part of Glasgow Railway system. Later, it became (65G), and eventually was just a diesel stabling point for various classes of shunters and Types 1, 2 and 3 locomotives. On 15th August 1979, when locomotives at this location were becoming less and less, two Class 37s, Nos. 37 156 and 37 152, lay parked in the open, amidst the barren waste remaining.

George O'Hara

GOUROCK S.P.

148 An Eastfield allocated Class 08 shunter could be found, in the mid 1970s, stabled at Gourock, on the Glasgow suburban electrified system. On 10th August 1976, No. 08 883 stood in the station, stabled beneath the wires. The £6¼ million scheme, to electrify the line from Glasgow Central to Wemyss Bay, via Gourock, commenced in 1965. Thirty-one road bridges and twenty-seven footbridges had to be increased in height over the route to accommodate the overhead wires.

Tom Noble

GRANGEMOUTH

149 This intriguing view of Grangemouth depot, with its track and pointwork snaking away from the covered accommodation, shows the variety of traction generally to be found at this location. Guarding the depot, on a Saturday in July 1978, are Class 20s, 27s, 37s and an 08 shunter.

Michael Rhodes

150 Grangemouth depot is ex-Caledonian Railway, and a good view of the structure of the shed building is seen in this view. The Caledonian Railway came to Grangemouth in 1866, when it was important to gain control of the thriving coal export trade to Europe from the Fife coalfields. Until 1866, Grangemouth with only two small docks belonging to the Forth & Clyde Navigation Co., was nothing more than a tidal harbour on the River Forth. To secure ownership of the area, it was necessary for the 'Caledonian' to buy up the complete dock area. The engine shed at Grangemouth was originally named Fouldubs. The area is now a busy industrial complex, with oil trains now also being a very common sight, as they travel to and from the oil terminal here. On 13th August 1978, Class 37, No. 37 147 stands at the depot, with Class 08 shunter, No. 08 246 and Class 20, No. 20 119 back in the depot.

Tom Noble

151 Lying back in the depths of the old steam depot at Grangemouth, lit by the shafts of light which penetrate through the shed roof, are two Class 20 locomotives, Nos. 20 121 and 20 101, coupled together, on 15th June 1981. The atmosphere of the days of steam always seems to remain on a visit to a location of this nature.

Rex Kennedy

152 The interior of Grangemouth depot, in this view, as seen on 9th May 1980, shows Class 20 locomotive detail of both nose and cab end, with Nos. 20 207 and 20 049 standing over the inspection pits. Two Class 37 diesels, Nos. 37 046 and 37 108 complete this interesting scene, which, to overcome the darkness within the shed, was photographed at one second exposure.

Antony Guppy

153 Class 20 locomotive, No. 20 115, stands beside the refuelling point at Grangemouth depot on 9th July 1979. The cab end on this particular locomotive differs from that of No. 20 207, seen in **Plate 152**, which carries the route indicator panel instead of the discs seen in this view. Two Class 08 shunters, Nos. 08 246 and 08 505 stand nearby. The main duties of the shunters are in the dock freight sidings at Grangemouth.

Tom Noble

154 Preparing to refuel at Grangemouth, on 15th June 1981, are two Class 27 locomotives, Nos. 27 022 and 27 102. This class of locomotive is a fairly common sight at this location, but as the depot itself only has a shunter allocation, most of the main line locomotives seen here, are Eastfield allocated.

Rex Kennedy

POLMADIE

155 Polmadie, with the old steam shed building, seen here, no longer in use, is a mere shadow of its former self. However, diesels were, for many years, prolific at the depot, as in this scene on 3rd August 1974. Classes on view include 17s, 20s, 24s, 27s, 37s, 40s and 47s. The vast expanse of the depot with its 14 roads is a formidable sight.
J. A. Sommerfield

156 Three Class 17 'Claytons' pose for a photograph outside the old steam depot at Polmadie, in April 1965. Steam locomotives are still in attendance on this occasion, and a water column can be seen beside one of the 'Clayton' locomotives.
David Thornley

157 Polmadie depot serves Glasgow Central station, and steam, diesel and electric locomotives, over the years, have used this location for coaling and refuelling, before returning to England over the West Coast Main Line. The 'Royal Scot' train has seen a variety of traction at its head, from LMS 4-6-2 Pacifics to Class 87 electric locomotives. On this occasion, 21st July 1961, Class 40, No. D297, carried the 'Royal Scot' headboard and stands on shed with steam locomotives simmering in the distance.
John Oatway

1P38
410

0S81

◀ **158** When the Class 50 locomotives were introduced on to British Rail in 1967, they were all allocated to the London Midland Region, and did not carry names until 1978, approximately three years after being transferred to the Western Region. Whilst allocated to the London Midland, they regularly worked as far north as Perth. On 15th November 1973, No. 410 stood in the sunshine in the yard at Polmadie.

Brian Morrison

160 Class 20s stand side by side at the depot entrance to Polmadie shed, on Sunday 11th August 1974. On this occasion, Nos. 20 096 and 20 223 stand over the pits, which remind us of the old steam days. The differing headcode systems are very apparent.

Rex Kennedy

161 A sight that has now disappeared, is that of the fine nameplates, with the shipping company standards, once carried by some of the early Class 40 locomotives. On 23rd March 1974, No. 40 015 *Aquitania*, one of the dual braked variety, lay amongst others of its class at Polmadie. These locomotives still, in fact, carry their names, but they are now only stencilled on the side of the locomotive.

Donald Cameron

162 After steam had disappeared, it was no longer important to differentiate between steam and diesel in the numbering system, by prefixing each diesel locomotive's number with the letter 'D'. Class 37, No. 6852, having had the 'D' removed, stands at Polmadie on 26th April 1973. Soon after this, however, the new T.O.P.S. numbering system was introduced.

John Oatway

◀ **159** Standing over the pits at Polmadie, where once stood ex-LMS 'Duchesses', 'Royal Scots' and 'Jubilees', Class 50, No. 50 034 waits on No. 12 road for her next call to duty, on 23rd March 1974, in the days when headcodes were still in operation.

Donald Cameron

163 One quite small building, erected at Polmadie, was fitted out for diesel maintenance. This type of maintenance shed, with its raised platforms, deep inspection pits and modern equipment, is, of course, commonplace at some of the major depots on British Rail, such as Tinsley, Toton and Cardiff (Canton). On 8th July 1978, Class 20s, Nos. 20 191 and 20 027 await attention in the company of Class 47, No. 47 211. Class 47s were not introduced to this area until 1965, when crew training commenced at Polmadie, using Class 47s, Nos. D1632, D1633, D1635 and D1639.

Michael Rhodes

164 Class 40, No. 40 060 stands in one of the maintenance bays at Polmadie, with its nose inspection hatch left open, on 11th August 1974. Although, on this occasion, this locomotive, amongst others at Polmadie, carried the new T.O.P.S. number, other locomotives, Class 08 shunters in particular, carried the old numbering system, including the 'D' prefix. *Rex Kennedy*

165 The depot area at Polmadie stands north of the main line from Glasgow (Central) station to Carlisle and the south. However, a line of Class 37 diesels, usually Motherwell allocated, can often be found here, south of the line. On 11th January 1979, No. 37 026 stands in the snow on a crisp winter's day. The diesel maintenance building appears in the background.

Tom Noble

166 The clock at Polmadie was a well-known feature of the depot, as were once, the Class 24 locomotives. The two classes of locomotive seen here on 15th June 1974, have now disappeared from the area, with No. 24 014, sadly withdrawn, and No. 50 021 transferred to the Western Region, and now bearing the name *Rodney*.

J. A. Sommerfield

167 Since the completion of 25 kV a.c. electrification to Glasgow (Central), the electric locomotive is a common sight at Polmadie. Set amidst the snow-bound area of the depot yard, Class 86, No. 86 018, stands beneath the wires on 11th January 1979. This particular electric locomotive has since been reclassified 86/3 and renumbered 86 318. The reclassification enables maximum speed to be increased from 80mph to 100mph, with the fitting of sab resilient wheels.

Tom Noble

168 The 'Condor', once labelled British Railways' fastest freight train, was an Anglo-Scottish container train, making the longest non-stop run of any train in Britain, covering 301 miles, overnight, between Hendon in London, and Carlisle. The train then continued on to the goods depot at Gushetfaulds, Glasgow, behind the same locomotive, a complete journey of 402¼ miles, with no change of motive power. This entire journey took under ten hours, giving an extremely reliable service to Scottish businesses. The Metrovick 1,200hp Co-Bo diesels worked this service from March 1959, and on 23rd May 1959, No. D5702 stood at Polmadie, still sporting its headboard, soon after arriving from Hendon. The 'Condor' service ceased before the withdrawal of the 'Metrovicks' from operating stock. No. D5705, bearing the number TDB 968006, still lives today, having been preserved by the Diesel Traction Group.

Norman Preedy

169 A fine example of a Class 06 shunter in the old livery with the 'lion' BR crest, in the days before yellow and black stripes were a feature of locomotive end design. No. D2428 stands, with wagons, at Polmadie on 16th August 1960. The Andrew Barclay works plate is situated above the number.

Peter Groom

170 One of the mainstays of Scottish Region motive power is the Class 26 locomotive. No. D5327, seen here, in 1968, at Polmadie, sports the green livery and yellow front end panel. On this locomotive, the front communicating doors stand ajar, and the single line token catcher is clearly visible beneath the cab windows, at the far end of the locomotive. Snowploughs are fitted to No. D5327, and the cover over the air horns at the front of the locomotive, on the roof, can be pulled down during snowstorms.

Norman Preedy

171 Disused Class 17 'Clayton' locomotives stand in the sunshine at Polmadie on 15th November 1973. After withdrawal, members of this class were to be found scattered about the Glasgow area, at various depots and at St. Rollox Works. Seen here are Nos. D8573, D8613, D8548 and D8504.
Brian Morrison

172 'Claytons', still during their working life, wait over the pits, outside the depot building at Polmadie, before returning to duty on 11th September 1971. Nos. D8552 and D8547 give an impression of power, with their two engines situated each side of the cab, but together they only provide 900bhp.
Norman Preedy

173 The diesel maintenance building, with shutters down, provides the background for two Class 08 shunters and a Class 40 locomotive at Polmadie on 27th April 1980. This now is virtually the only spot at Polmadie where diesels can be found, except on the other side of the main line. No. 40 151 stands at the buffers, and No. 08 754 is seen together with No. PO. 1, a shunter, which once carried the number 08 173 when in BR operating stock. No. PO. 1 is now used for shunting duties at the regional repair shops at Polmadie.

Rex Kennedy

174 This new building at Polmadie will eventually house the 25 kV overhead electric Advanced Passenger Train, when it comes into regular service from Glasgow to London. This fine structure is in vast contrast to the old Caledonian Railway steam shed design, and should allow the entire train to be accommodated. The building itself will only be used for occasional maintenance and the cleaning of the APT and Mk III coaching stock, as general maintenance of the APT will be carried out at Glasgow (Shields Road). On 27th April 1980, a solitary Class 81 electric locomotive, No. 81 003, stood outside this depot.

Rex Kennedy

GLASGOW (Shields Road)

175 A crisp winter day at Glasgow (Shields Road), showing the large steel structure built to house electric locomotives, multiple units and also the Advanced Passenger Train. On 11th January 1979, Class 86 electric locomotive, No. 86 010, now reclassified and renumbered 86 310, stands at the depot entrance.

Tom Noble

176 A look to the future. A Class 370 Advanced Passenger Train, No. 370 002, protrudes from the unusually designed shed at Glasgow (Shields Road) on 19th August 1979. These overhead electric trains comprise six articulated trailer cars, plus one power car in the centre of the train. However, they will normally double up with the two power cars coupled together and six trailer cars either side, when in service. In fact, set No. 370 002 paired with set No. 370 006 formed the first scheduled service, which left Glasgow (Central) on the morning of Monday, 7th December 1981, and took just under 4¼ hours to complete the journey to London (Euston); a fine achievement.

Tom Noble

177 Inside the three road shed at Glasgow (Shields Road) stands Class 81 electric locomotive, No. 81 016, sandwiched between APTs which eventually will regularly work over routes now worked by the Class 81–87 electric locomotives. This particularly interesting view, on 27th April 1980, shows APT sets Nos. 370 004 and 370 006 in different modes. On the right, we see the APT, with the front raised, looking like something from outer space; an unusual view. The Class 81 was one of the first main line electric locomotives introduced on to British Railways in 1959. The vast change in design over a twenty year period is very apparent.

Rex Kennedy

178 On 15th September 1978, an 'open day' was held at Glasgow (Shields Road). Locomotives which did not often appear here were on view. This fine portrait of Class 20, No. 20 137, in ex-works livery, was taken on this occasion.

Tom Noble

179 Lines of electric multiple units are to be found at Glasgow (Shields Road), these being mainly Class 303 sets. All are designed with air-operated sliding doors, and were introduced in 1960. Nos. 053 and 036, now bearing the 303 prefix to its number, lie here on 19th August 1979. An APT can be seen in the background.

Tom Noble

180 Class 303, Glasgow area, three-car electric units, stand inside the depot at Shields Road, and catch the sunlight streaming in through the doors on 15th September 1978. Class 303 and 311 sets, allocated to Shields Road, generally operate on Glasgow suburban routes, south of the Clyde, the 311s being used over Glasgow (Central), Gourock, Wemyss Bay and Cathcart Circle routes. None of the Class 314 units are allocated to Shields Road.

Tom Noble

181 The Scottish Region Class 920 prototype three-car electric unit, No. 920 001, is pictured at Glasgow (Shields Road) on 27th December 1977. The driving car at the front has an all metal finish, whereas the other two cars in the set are painted blue. It was built in 1975 with air-operated sliding doors and gangwayed throughout, and, in design, was the forerunner of the Class 314 and 315 sets, now to be found running on the Scottish and Eastern Regions. This prototype set can now be found at the Railway Technical Centre, Derby.

Tom Noble

182 An invader from the East Coast Main Line, in the form of HST set No. 254 003, takes its place at the 'open day' at Glasgow (Shields Road) on 15th September 1978. Standing alongside, with pantograph raised, is Class 87 electric locomotive, No. 87 002 *Royal Sovereign*, heading car transporter wagons.

Tom Noble

183 The streamlined designs of both diesel powered HST set No. 254 007 and electrically powered APT set No. 370 001, are very apparent in this front end view, taken at Glasgow (Shields Road) on 15th September 1978. The APT appear as British Rail's answer to the 'Concorde', and the HST design is similar to that of the new XPT train, which runs with only five trailer coaches in New South Wales, Australia.

Tom Noble

HYNDLAND

184 Hyndland e.m.u. depot basically serves the Glasgow suburban routes north of the Clyde, with Class 303 and 314 units allocated to this depot. The Class 314 units were built in 1979 for use on the new electrified Argyle line from Partick to Rutherglen. This necessitated electrifying an already existing route, and the re-opening of tunnels into Glasgow (Central), Low Level station. On 22nd August 1979, unit No. 314 212 stands in Hyndland yard, and, when photographed, had not entered service. Also in this view, we see Class 131 single car d.m.u. No. SC55014, now withdrawn, with Class 311 e.m.u. No. 311 092; two unusual visitors to Hyndland.

Tom Noble

185 A winter scene at Hyndland depot. Class 303, Glasgow area, three-car e.m.u. No. 303 025, built in 1960, stands beneath the depot wires in its freshly applied blue and grey livery, on 9th January 1979.

Tom Noble

BRIDGETON CENTRAL S.P.

186 Glasgow area, three-car e.m.u. sets lie stabled at the buffers at Bridgeton Central depot on 7th September 1979. As can be seen from the buckets and mops, cleaning is carried out at this location. All sets are sporting the smart blue and grey livery.

Tom Noble

BALLOCH CENTRAL S.P.

187 A Class 303 e.m.u. set can often be found stabled at Balloch Central. From here, the line continues north for a short distance to the terminus at Balloch Pier, where the Loch Lomond steamer, owned by Caledonian MacBrayne, can be boarded. On 11th April 1980, No. 303 025 waits to return to duty.

Les Nixon

AIRDRIE S.P.

188 Hyndland allocated Class 303 electric units lie stabled against the broken wooden buffers in Airdrie sidings, at the end of one of the Glasgow suburban electrified routes. At this point, a freight-only line continues east to Edinburgh, joining the Edinburgh (Waverley) to Glasgow (Queen Street) route at Bathgate Junction. Class 303s on view, on 25th January 1981, were Nos. 303 047, 303 016 and 303 052, awaiting return to duty the following morning. These three units were some of the many built in 1960. As every other locomotive and unit at that time was painted green, with the exception of these Scottish units, they attracted the name 'Blue Trains'. The introduction of the 'Blue Trains' on Clydeside trebled revenue north of the Clyde in the first three years from their inception, on 7th November 1960, resulting in the extension of the system south of the Clyde.

Brian Morrison

HELENSBURGH CENTRAL S.P.

189 Helensburgh has two stations, 'Central', a terminus station pictured here, and 'Upper', on the West Highland line to Fort William and Oban. The electrified Glasgow suburban route terminates at Helensburgh Central, and units are regularly stabled here between duties. On Sunday, 25th January 1981, various Class 303 units were stabled, both inside and outside the station.

Brian Morrison

MILNGAVIE S.P.

190 The winter sunshine gleams through the heavy clouds at Milngavie (which is pronounced Milguy), on 26th January 1981. Two Class 303 electric units, Nos. 303 046 and 303 034 lie stabled, whilst No. 303 047 forms the 14.55hrs. train to Airdrie. The British Rail symbol, seen clearly outside the station, is based on the idea of two-way traffic movement.

Brian Morrison

191 Motherwell depot is situated just off the main Glasgow (Central) to Carstairs electrified line, and locomotives use the old steam depot built by the Caledonian Railway. The building now houses locomotives in one part and wagons in another. On 14th June 1981, two Class 37 locomotives, Nos. 37 133 and 37 148 stand outside the depot, with a Class 20 inside.

Rex Kennedy

192 One of the few Class 47 locomotives fitted with snow-ploughs, No. 47 464, stands at the entrance to Motherwell depot, near the wagon shops, on 14th June 1981. It is noticeable that, whereas in the previous view, the building has a complete brick frontage, in this picture, showing the other end of the shed, the sections to the roof apex have been replaced by steel panels.

Rex Kennedy

193 A fine study of Class 17 'Clayton' locomotive, No. D8505, as it rests in the sunshine at Motherwell depot on 18th May 1964. In November 1967, the earlier members of this class could be found standing in line at the old steam depot at Carlisle (Kingmoor), amidst the 4-6-2 'Britannias' that were all allocated there at that time, the 'Claytons' having recently been transferred there from Polmadie. Kingmoor depot eventually became their last home prior to withdrawal.

Peter Groom

194 On 17th August 1969, Class 17 'Clayton' locomotive, No. D8567, stood where, as late as 1966, ex-LMS 'Black 5s', 'Crabs' and even ex-LNER B1s held their place, together with BR standard locomotives. To the right of the picture a glimpse of Class 47, No. 1111, which was the last Class 47 built at Crewe Works, can be seen.

Norman Preedy

195 Beneath the footbridge known to all who visit Motherwell depot, a Class 08 shunter, No. 08 732, lies stabled on 13th January 1980. The shunters allocated to this depot are used in the Lanarkshire yards and steelworks in the vicinity.

Paul Judge

196 A part of the old steam depot at Motherwell has been converted into a modern diesel maintenance building, with raised walkways of the kind found at diesel depots throughout Britain. This modern fitted out building constantly houses a locomotive under repair, and Class 20, No. 20 115 is seen here on 19th August 1979. The Class 20s found at Motherwell are used for internal Scottish freights, ballast trains and m.g.r. workings from Killoch and Barony to Longannet Power Station.

Tom Noble

197 Quite a variety of main line traction is to be found at Motherwell, including Class 20s, 25s, 37s, 40s and 47s. Class 25, No. 25 140, is one of the batch with no front gangway, a roof headcode box and roof-mounted air horns, and was pictured at this location on 6th June 1978.

Norman Preedy

198 Two English Electric Class 37 locomotives, Nos. 37 155 and 37 151, stand beside the depot building at Motherwell on 21st September 1978. The Class 37s here, work the Hunterston to Ravenscraig steelworks ore and coal trains, double-headed with regular daily workings, and they are also used on some internal Scottish freights. Some of the Class 37s allocated to Motherwell are now sporting silver painted roofs.
Tom Noble

199 With a clear view through the old steam depot building at Motherwell, and sight of the section now converted to a diesel maintenance workshop, Class 40, No. 40 169, stands beside the fuel tanks adjacent to the maintenance shed, together with Class 47, No. 47 562 on 14th June 1981. The fittings between the buffers on the Class 40 locomotive indicate that it is fitted for multiple working.
Rex Kennedy

200 Class 37 locomotives are a familiar sight at Motherwell depot. On 2nd January 1978, No. 37 144 stands alone in the sunlight outside the depot building, showing clearly the design of the structure.

Tom Noble

201 Class 40 locomotives are also regular visitors to Motherwell, although Motherwell does not actually have an allocation. Beside the depot building and beneath the footbridge stands No. 40 101, on 3rd November 1979.

Graham Scott-Lowe

203 The interior design of the d.m.u. maintenance depot at Hamilton, is in keeping with the style of building used for this purpose throughout the British Rail network. Elevated trackwork and continuous maintenance pits throughout the building form the scene, as Metro-Cammell unit, Class 102/1, headed by power car No. SC51805, sits over the pits, together with Class 116/2 Derby Works BR three-car suburban unit, with power car No. SC50841 in view, on 21st September 1978.

Tom Noble

HAMILTON

202 The facade of the corrugated iron and steel shed building, which houses multiple units at Hamilton, is clearly illustrated in this view taken on 12th August 1979. On this occasion, all lines were occupied by Class 107 Derby Works B R three-car units and a solitary Class 131 Gloucester R. C. & W. Co. single-car unit.

David Maxey

204 Derby Works B R Class 107/2 unit, headed by motor brake second car No. SC52000, starts her engines beside Hamilton depot, on a crisp winter day in February 1978. Units allocated to Hamilton, operate services from Glasgow (Central) to the west of Scotland, with the exception of the majority of Ayrshire services.

Tom Noble

205 Bearing the Trans-Clyde Greater Glasgow crest on the side of the unit to the left of this picture, a Class 107 Derby Works B R unit carries the name 'Largs' in the destination panel. Largs is one of the only locations in Ayrshire served by Hamilton units. Together with another member of this class of d.m.u., she stands over the pits, inside Hamilton depot, on 13th January 1980.

Paul Judge

206 The sight of one of these fascinating machines in Scotland was one not to be missed in the early 1960s. Only two of these Bristol ECW four-wheel railbuses were built and carried the numbers SC79958 and SC79959. On 20th July 1961, the one in this picture could be found on Hamilton depot. These rail-buses provided seating accommodation for 56 passengers.

John Oatway

Shunters at Hamilton

207 An interesting view of a Hamilton allocated Class 08 shunter, No. 08 421, standing outside one of the smaller buildings on its home depot. The locomotive is at rest, on Sunday 22nd March 1981, coupled to wagons containing d.m.u. engines returned from major overhaul.

Tom Noble

208 Since the sad withdrawal of the Class 06 shunters from the Scottish Region, the shunter scene is now entirely Class 08. However, in the 1960s and 1970s, a Class 06 shunter could regularly be seen at Hamilton depot. On 12th June 1975, No. 06 004 stood in the depot yard, bearing an Eastfield (ED) allocation code on the cab side.

Norman Preedy

209 On 8th August 1966, Class 06 'Barclay' shunter, No. D2410, was seen at Hamilton depot. The change in style of the actual numerals is apparent when compared with those portrayed in the previous picture. This particular locomotive was withdrawn before the T.O.P.S. numbering system was introduced.

David Wood

MOSSEND YARD S.P.

210 Mossend Yard, just south of Whifflet Junction and east of Glasgow, is the point where electrically hauled trains, from Carlisle and the south, change motive power and become diesel hauled, before continuing to Perth, Aberdeen and Inverness. Just north of this location, the overhead electric wires cease to exist, and often large quantities of electric locomotives can be found stored at Mossend, pending return to duty with south bound traffic. On 21st September 1978, in this view of the Mossend location, looking north, a variety of Bo-Bo electrics stand here, headed by Class 85, No. 85 017 and Class 87, No. 87 006 *City of Glasgow.*

Tom Noble

211 This profile, taken on 14th June 1981, shows three classes of electric locomotive, stabled beneath the wires at Mossend Yard, and Nos. 81 021, 85 031 and 86 037 are pictured. The change in style of the front end of the Class 86, in comparison to that of the Class 81 and Class 85 is clearly visible. The Class 81 Bo-Bo electrics were the first class of 25 kV a.c. electric locomotives to be built in 1950, and were originally fitted with two operating pantographs, although today, they only carry one.

Rex Kennedy

CARSTAIRS

212 Carstairs is situated on the main Glasgow to Carlisle electrified line, at the point where the line joins from Edinburgh. Trains from the south often split at Carstairs, the second half of the train being taken over the non-electrified section to Edinburgh, usually hauled by a Class 47 locomotive. The same situation, of course, happens in reverse, where the two trains join at Carstairs before continuing south. This creates the situation of both diesel and electric traction being stabled at this location, and on 9th July 1977, the scene from the station not only shows three different classes of diesel and electric power, but also the test coach *Hermes*.

Donald Cameron

213 Bearing the shed code 62B, and its manufacturer's name on the front, Class 06 shunter, No. 06 010, previously numbered D2444, lies at rest inside the depot building at Carstairs on 16th April 1977. Alongside is seen another, now withdrawn, locomotive, Class 24, No. 24 128.

Paul Brinsford

214 This closer view of the depot area at Carstairs, shows Class 47/7 diesel No. 47 703 *Saint Mungo*, a class, although fitted for push-pull working on the Edinburgh (Waverley) to Glasgow (Queen Street) line, is often used on trains from Carstairs to Edinburgh. Beside her stands Class 86 electric locomotive, No. 86 039, with the depot building in the background, on 27th August 1979.

Donald Cameron

215 A rare sight at Carstairs was that of Class 55 'Deltic', No. 55 014 *The Duke of Wellington's Regiment* on 18th August 1979. The 'Deltic' is seen here passing the station on its way to the depot yard seen in the background, with the shed building, now since dismantled, still standing. Scottish trials for the Class 55 'Deltic' locomotive took place on 6th June 1959, the first trial run following the Waverley Route from Carlisle, eastbound, eventually terminating at Leith Central.

Donald Cameron

BEATTOCK

216 For many years, two Class 20 locomotives were stabled south of Beattock station, for banking purposes. The two stabled, awaiting their next call to duty, on 4th August 1978, are Nos. 20 086 and 20 114. Locomotives are no longer stabled here for this task. The last steam locomotives to work this duty were 2-6-0 Class 4 Standards and 2-6-4 BR Standard tank engines, in 1966. The summit (1,015 ft) lies ten miles north of Beattock station, and from it, the main line descends north into the Clyde Valley.

Graham Scott-Lowe

217 The three-road modern structure used to house diesel multiple units at Corkerhill, seen in this view, shows its sizeable width, and raised walkways which pass through it. Three-car Derby Works B R Class 107 units fill the depot building on 27th April 1980, whilst a Metro-Cammell Class 101 unit stands alongside. All the units on view carry the Greater Glasgow Trans-Clyde emblem.

Rex Kennedy

218 The two fuelling bays at Corkerhill, seen in this picture, stand beside the 'Canal' line, the route from Glasgow (Central) to Paisley (Canal). On 21st October 1979, Class 45 'Peak', No. 45 011 stood in the fuelling bay, which lies nearest to the depot complex, whilst Class 47, No. 47 543 could be seen in the foreground. In the distance is the old Glasgow and South Western signal box, which stands beside the 'Canal' line, a line scheduled for closure in January 1983, should there be no improvement in passenger receipts over this route.

Tom Noble

219 On a bright, but cloudy, day at Corkerhill, a variety of multiple units bask in the sunshine, in the depot yard, on 15th September 1978. This scene portrays Class 107/1 Derby Works B R unit, headed by car SC52017, together with Metro-Cammell Class 102/1 set, showing car No. SC51804. Resting in the shed building is another Metro-Cammell set, this time a Class 101/1, with car No. SC50175 lit by the sunlight. Since this picture was taken, all Class 102/1 units have been reclassified Class 101. Although Corkerhill has no allocation of its own, Hamilton units are often serviced at this location. The unusual lighting gantries are a feature of this depot.

Tom Noble

220 The new Class 140 two-car experimental d.m.u. No. 140 001, spent three weeks in the Glasgow area during August 1981, and is pictured here, in Corkerhill maintenance depot, on 28th August 1981. Each car of the Class 140 unit is mounted on four wheels, as were the early railbuses, of the type shown in **Plate 206**, and the two-car set has been tried over many routes throughout Britain. Whilst in Glasgow, it spent its first week operating the Cumbernauld to Springburn service, and the remaining two weeks on services from Glasgow (Central) to Barrhead, Kilmarnock and East Kilbride. This view shows the unit undergoing inspection in a building which was extended in 1981.

Tom Noble

221 In 1969, Class 17 'Clayton' locomotives were still in use on the Scottish Region, and No. D8577 is pictured at Corkerhill in August of that year. From the late 1960s to the present day, new modern buildings have sprung up at Corkerhill, creating an ultra modern diesel depot complex, replacing the old steam sheds, now demolished. At the end of steam in Scotland, around 1966, ex-LMS 'Black 5s', Class 4 and Class 5 BR Standard 4-6-0 locomotives and BR Standard 2-6-4 tank engines made up the depot allocation.

Roy Jones

222 Corkerhill was originally a large railway community which developed in 1896. It consisted of a steam depot, with terraced railway cottages and a railway institute in close proximity. The development at Corkerhill was felt necessary owing to the congestion being created at Glasgow (St. Enoch) depot. The cottages were demolished in 1971, and have now been replaced by blocks of flats, but the institute still stands today. During 1961, steam and diesel traction could be seen together at this depot, and in this view, taken on 5th May 1961, Class 45 'Peak', No. D24, stands beside the new office block, whilst in the background lies ex-LMS 4-6-0 'Jubilee' No. 45727 *Inflexible*, soon to be withdrawn.

Bill Smith

223 On 26th April 1973, Class 40 locomotive No. 269 stood at Corkerhill, outside the, now deteriorating, old steam shed buildings. The pits remain, but part of the roof has gone. The last of the old buildings which formed the steam depot at Corkerhill have now been demolished, and the actual windows from one of the shed buildings can be found in the locomotive shed built by the Scottish Railway Preservation Society at Bo'ness in West Lothian.

John Oatway

224 Class 06 'Barclay' shunter, No. D2440, prepares to continue its duties at Corkerhill on 21st July 1961. These locomotives carried their number between the front buffers, as seen in this view. After T.O.P.S. renumbering, this locomotive became No. 06 009, and at that time was allocated to Dunfermline (Townhill).

John Oatway

AYR

225 The depot at Ayr is situated north of the station and south of Falkland Yard, beside the main line from Glasgow to Stranraer. Many new modern style buildings have been constructed at this location, and two differing designs are seen in this view. On 2nd June 1977, Class 20 locomotives, Nos. 20 137 and 20 099 stand in the yard, together with Ayr allocated Class 08 shunter, No. 08 344. A Swindon Works BR Class 126/1 unit, showing car No. SC51013, protrudes from the shed building.

Brian Morrison

226 Ayr motive power depot, when built by the Glasgow & South Western Railway, was situated in a triangle formed by the Ayr/Newton on Ayr/Auchincruive lines. Unfortunately, the line which bypassed the depot on the east side, from Ayr to Auchincruive, has now gone, and the only passenger carrying route existing today is the one which passes between the depot and the coast, from north to south. The present depot, although being the resting place for a variety of locomotive classes, only has allocated, in addition to diesel multiple units, Class 08 shunters. One of these, No. 08 433, is seen in this view on 26th June 1978, with two Class 20s, Nos. 20 115 and 20 112 at the fuelling point nearby. Shunter duties at Ayr include Falkland Yard, Ayr and Ardrossan Harbours and Irvine permanent way depot.

Tom Noble

227 Some of the old stone buildings, built by the Glasgow & South Western Railway, still stand today, and in fact are still in use for housing locomotives. This scene, captured in the summer of 1973, shows Class 20, No. 8321, standing outside one of the original buildings, with Class 24, No. 5009 just inside. As a point of interest, the first Class 20 locomotive was handed over to British Railways on 3rd June 1957.

Richard Charlson

228 We constantly seem to be seeing a variation in roof and front end paint designs on locomotives throughout the British Rail network. Stratford depot, on the Eastern Region, together with Eastfield and Motherwell on the Scottish, are perhaps setting a trend. The majority of the 'new look' locomotives are rather attractive, and on 31st August 1981, Class 20, No. 20 122 of Motherwell depot, stood near the fuel tanks at Ayr depot, proudly displaying her new livery.

John Hillmer

Refuelling

229 The two road fuelling point at Ayr depot is rather unusual in design, and this picture clearly portrays its features. Awaiting refuelling from the centrally situated fuelling equipment, Class 27 locomotive, No. 27 210, prepares for its next duty on 31st August 1981. This particular locomotive was one of the batch of twelve converted from Class 27/1 to Class 27/2, and fitted with electric train heating.

John Hillmer

230 Another fuelling point at Ayr depot is one of the standard design found at many depots throughout Britain. On 16th August 1976, Class 25/0 locomotive, No. 25 009, stood taking on fuel. The omission of the roof mounted horns is clearly noticeable, differing in this feature from that of the Class 25/1 locomotives of Motherwell, as shown in **Plate 197.**

Michael Rhodes

231 A view of Ayr depot on 15th August 1967, showing clearly the new modern steel structure, built in 1960, to house diesel locomotives, beside the old Glasgow & South Western steam shed. In fact, two steel extensions were built at each end of one of the original stone shed buildings, and inside this stone building, which now forms the centre section of the new depot, the pillars and pits from the steam days still remain. A Class 06 shunter, sporting its front number, stands in the new depot together with a d.m.u., whilst Class 20 and 25 locomotives lie at rest where steam once held pride of place. The Class 20s are used on trains to and from the Ayrshire coalfields.

John Hooper

232 When steam engines lay dying at Ayr depot on 7th January 1967, a Class 05 'Hunslet' 0-6-0 shunter, carrying the departmental number 88, stood in the depot yard, devoid of its side panels, and exposing the intricate workings of its diesel engine. The original number of this locomotive was D2612, and it was built in July 1960. It entered operating stock at Darlington (51A) in January 1961, and during that year it became departmental locomotive No. 88 and was used at Faverdale Works, Darlington, who, at that time, were building the SPVs, (special parcels vans), for British Railways. It was 1967 before its allocation was changed to Ayr and part of its duties were at Barrassie Wagon Works.

Roger Lamb

233 Near the familiar lattice iron footbridge, which crosses the line at Ayr depot, lies Class 06 'Barclay' shunter, No. D2443, on 5th October 1965. About this time, Ayr depot had about three members of this class included in its locomotive allocation.

Barry Wynne

234 A sight from the past is that of the coaling tower at Ayr, beneath which many a fine ex-LMS 'Crab' or 'Black 5' once stood awaiting fuel for its fire. On 7th January 1967, a Class 24 diesel locomotive occupied this position.

Roger Lamb

235 This scene, inside the d.m.u. maintenance depot at Ayr, shows a sight which was commonplace over the Ayrshire railway routes. Peculiar to the area, until late 1981, were the Swindon Works BR Class 126 Inter City units, with blocked-off corridor connections at the end of the train. Two of these unusual sets stand over the inspection pits at Ayr depot on 27th December 1979.

John Hillmer

IRVINE S.P.

236 On 19th September 1979, Class 08 shunter, No. 08 730, lay stabled in the sunshine at Irvine station. The locomotive, allocated to Ayr depot, is used at the permanent way sidings nearby. Irvine is situated on the main line between Ayr and Kilwinning, where the line joins from Largs.

Tom Noble

STRANRAER S.P.

ARDROSSAN

238 Ardrossan was originally served by both the Caledonian Railway and the Glasgow and South Western Railway, but the shed building, shown here, is of G&SW design. It was situated in the triangle formed by the Glasgow-Ardrossan, Glasgow-Largs and Ardrossan-Largs, ex-G&SW Rly lines. In September 1972, Class 17 'Clayton' locomotives were stored at this location, inside and outside of the depot building which has since disappeared.

Bill Smith

▼ **239** Another view of the structure of the depot at Ardrossan, on 25th August 1968. The remains of the water column are still visible in this scene looking through the depot. A Cravens d.m.u. stands protruding from the depot entrance, and Class 08 shunter, No. D3561, can just be seen inside the shed alongside this unit. *Roger Lamb*

237 Stranraer once boasted two fine stations, the 'Town' and the 'Harbour'. Now, alas, only the Stranraer Harbour station is used. However, the old steam depot, where, in recent years, diesels have been stabled, was situated near the 'Town' station. This view, on 30th March 1980, portrays this location with Class 25, No. 25 238, stabled together with Class 08 shunter, No. 08 730. Swindon Works BR Class 126 multiple unit set showing car SC51013 with blocked-off corridor connection, completes the scene. The sight of an 08 shunter at Stranraer is not a familiar one. Two routes once converged on Stranraer; from the east came the line from Dumfries via Newton Stewart, sadly closed on 14th June 1965, and from the north is the existing route from Glasgow via Ayr.

Tom Noble

ARDROSSAN HARBOUR

240 Today, Class 08 shunters used at Ardrossan are to be found in the harbour area. These shunters are officially allocated to Ayr, and on 20th January 1980, Nos. 08 344 and 08 433 lay stabled in the harbour sidings. The code 67C is plainly visible on the rear of the shunter in the foreground, a code that was discontinued in 1973, the new code being AY.

Brian Morrison

DUMFRIES

241 Dumfries once had a fine six-road steam depot built by the Glasgow & South Western Railway, and in the mid 1960s, this was being extensively used by diesel locomotives working in the area. On 7th August 1966, Class 27, No. D5362, now 27 016, was seen outside one of the six portals, which made this shed building rather unique. The locomotive is sporting the original BR green livery with a continuous white horizontal line around the body. Snowploughs, a common feature on Scottish allocated locomotives, are also fitted.

Roger Lamb

242 This view of the interior of the shed building at Dumfries, emphasises the vast width of the depot, with its inspection pits, pillars and fine roof structure. The scene, on 7th August 1966, includes Class 25 locomotives, Nos. D7611 and D7615, with Class 08 shunter, No. D3926, on the far road. The last steam locomotive allocated to Dumfries in 1966 was ex-LMS 'Black 5' No. 45480.

Roger Lamb

243 Since the demolition of the shed building at Dumfries, locomotives are stabled in the bay platform at Dumfries station. The station is a stone structure, with a ridge and furrow awning and the pillars to the roof bear the G&SW motif. The Class 08 shunter now operating at Dumfries, is allocated to Carlisle (Kingmoor), and is the only shunter working in Scotland which is maintained by a London Midland Region depot. Class 25, No. 25 240 was seen in the bay on 23rd September 1979, together with Class 08 shunter, No. 08 452.

Tom Noble